INTRODUCTION

The history of electromagnetic (EM) theory begins with ancient measures to understand atmospheric electricity, especially lightning. People then had little understanding of electricity, and were unable to explain the phenomena. Scientific understanding into the nature of electromagnetism grew throughout the eighteenth and twentieth centuries through the work of researchers such as Ampère, Coulomb, Oersted, Faraday, Franklin,

Maxwell, Heaviside, Tesla, Meyl, Bearden and many others. Because of centuries of scientific experimentation and continual advancement of knowledge, our lives have been considerably enhanced by the supply of electrical energy to power our homes, lives, and industries. Yet with every scientific advancement, there is the potential for good and evil applications for its use, depending on the morality and intents of the users.

A good man brings good things out of the good stored up in his heart, and an evil man brings evil things out of the evil stored up in his heart. For the mouth speaks what the heart is full of. Et godt menneske henter gode ting frem af sit hjertes gode forråd. Et ondt menneske henter onde ting frem af sit hjertes onde forråd. For hvad hjertet er fuldt af, løber munden over med. Luke 6:45

Anyone who has experienced an electrical shock is aware of the obvious dangers of electricity, but the subtler uses of EM are not as obvious (electromagnetic pulses (EMP) which can destroy computer networks, mind control and control of individuals' behaviors, scalar EG (electrogravitational) beams and "death rays", weather and climate modification, and disruption and destruction of the electrical power grids of the world.

Enhver, der har oplevet et elektrisk stød er klar over de åbenlyse farer ved elektricitet, men de finere anvendelser af EM er ikke så indlysende (elektromagnetiske impulser (EMP), som kan ødelægge computernetværk, tankekontrol og kontrol af den enkeltes adfærd, skalar EG (electrogravitational) stråler, vejr og klima ændringer, og forstyrrelser og ødelæggelse af de elektriske elnet i verden.

This book will discuss a wide range of beneficial uses of electromagnetism (medical procedures, diathermy, supply of relatively inexpensive and "free" energy) as well as some of the dangerous and destructive technologies which threaten human existence on this planet, such as scalar weapons, electromagnetic warfare, and electromagnetic biological (EM BM) warfare.

Robert Oppenheimer once warned us about the "sin against humanity" of physicists in employing weapons of mass destruction against God's creation. "Despite the vision and the far-seeing wisdom of our wartime heads of state, the physicists felt a peculiarly intimate responsibility for suggesting, for supporting, and in the end, in large measure, for achieving the realization of atomic weapons. Nor can we forget that these weapons, as they were in fact used, dramatized so mercilessly the inhumanity and evil of modern war. In some sort of crude sense which no vulgarity, no humor, no overstatement can quite extinguish, the physicists have known sin; and this is a knowledge which they cannot lose. There are profound implications regarding the use of electromagnetic energy to destroy human life, and our choice of world leaders can affect humanity's future for good or evil!

In the 1930s, the Soviets obtained some of Nikola Tesla's scientific documents, while the West ignored his findings. By 1981 the Soviet Union had discovered and weaponized the Tesla scalar wave effects. There is also evidence, mostly classified, that the spy plane piloted by Gary Powers in 1960 was destroyed by these weapons. The most advanced US nuclear submarine, the U.S.S. Thresher, was also sent to a watery grave in 1963, with 129 American sailors, using scalar weaponry. There are several Tesla devices of note, but the most powerful of these frightening weapons is the Tesla Howitzer. It was completed at the Saryshagan missile range and presently considered to be either a high-energy laser or a particle beam weapon.

Today, in the 21st Century, Britain, China, Russia and the United States are actively continuing research in scalar weaponry, but the United States is slightly behind the race. Scientists in the West ignored Telsa in the 1930s, while his research was stolen by Soviet and German agents.

The key concept is that scalar waves restore certain useful aspects of Maxwell's equations, which were omitted by Oliver Heaviside. The scientific evolution of classical electrodynamics was "flawed" due to the omissions in the 19th century by Heaviside, Hertz and Gibbs. The simplification of Maxwell's equations by deleting the magnetic flux contributions to energy fields has had widespread implications for the use of scalar devices to improve the quality of human life on the planet, or to destroy it!

Russian energetics scientists corrected the electromagnetic models, and went to develop the Classified Unified Field Theory. This led to the development of the powerful scalar interferometer superweapons. The Soviets in 1985 once threatened the earth itself by activating their scalar weapons with multiple scalar transmitters turned on simultaneously, endangering the survival of the entire planet. Per nuclear physicist Thomas Bearden, they conducted a massive, 'full up' scalar weapon systems and communications strategic exercise. During this exercise, American Frank Golden discovered that the Russians had activated 27 gigantic "'power taps." These were established by resonating the earth electrogravitationally on 54 powerful scalar frequencies (27 pairs where the two are separated from each other by 12 kHz.), transmitted into the earth. The Soviets utilized this to stimulate the earth into forced electrogravitational resonance on all 54 frequencies. Each of the 27 power taps extracted enormous energy from the molten core of the earth itself, and turning it into ordinary electrical power. Each giant tap was capable of powering 4 to 6 of the largest scalar EM howitzers possessed by Russia. Bearden wrote: "Apparently over 100 giant scalar EM weapons were activated and a large number of command and control transmissions and it lasted several days. By alternating the potentials and loads of each of the two paired transmitters, electrical energy in enormous amounts can be extracted from the earth itself, fed by the 'giant cathode' that is the earth's molten core. Scalar EM command and control systems, including high data rate communications with underwater submarines, were also activated on a massive scale. The exercise went on for several days, as power taps were switched in and out, and command and control systems went up and down. Bearden claims not one American intelligence lab, or scientist detected this as they didn't have a detector for scalar EM radiation, and that not one officially believes that the exercise ever happened." However, it was monitored on an advanced, proprietary detection system by Frank Golden for several days and by Bearden for several hours.

A secondary purpose of this book is to integrate moral considerations into the decision to promote and develop potentially dangerous technologies. An informed and intelligent citizenry has the right to challenge the use of destructive technologies and to prevent their employment. This citizenry does not exist, and is kept in the dark by the governments of the world.

Matthew 12:35
A good man out **of** the good treasure **of** the **heart** bringeth forth good things: and an evil man out **of** the evil treasure bringeth forth evil things.

There are also many translations of source documents in several foreign languages, which allow my language students to gain a technical and scientific vocabulary, and to be able to read some original European dissertations and research studies.

1. THE SOURCE OF ELECTROMAGNETIC RADIATION

The sun radiates energy equally in all directions, and the Earth intercepts and receives part of this energy. The power flux reaching the top of the Earth's atmosphere is about 1400 Watts/m². What this means in practical terms is that, on an average, one square meter on the side of the Earth facing the sun receives energy from the sun equal to that from approximately fourteen 100 Watt light bulbs every second! It can be seen from the chart of the electromagnetic spectrum that this visible light passes through the atmosphere.

THE ELECTROMAGNETIC SPECTRUM

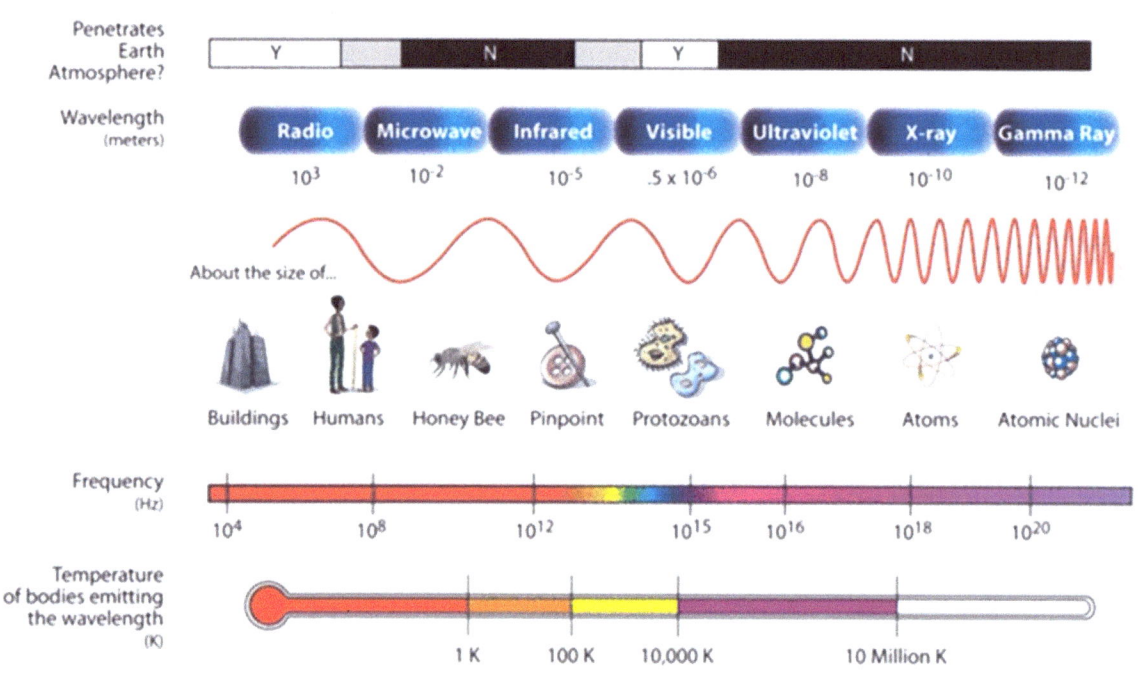

2. ELECTROMAGNETICS – FROM HANS CHRISTIAN ØRSTED TO NIKOLA TESLA, KONSTATIN MEYL, THOMAS BEARDEN, AND MANY OTHERS.

Continual progress has been made in science, with many contributors. Max Planck observed that "It was not by any accident that the greatest thinkers of all ages were deeply religious souls." Since the devastation of Hiroshima and Nagasaki by the atomic bomb in 1945, there seems to be little restriction of the use of weapons of mass destruction (WMD) in warfare.

Some of the greatest scientific discoveries in history are flawed by false assumptions, oversimplification of complex relationships, reductionism, and the omission of significant contribution factors. This includes such great scientific contributors to electromagnet theory and practice as Albert Einstein, Nicola Tesla, Oliver Heaviside, and Josiah Willard Gibbs.

Benjamin Franklin investigated electricity and tested theories with an extremely dangerous experiment of having his son fly a kite through a storm-threatened sky. A key attached to the kite string sparked and charged a Leyden jar, which stored electrical charge. Thus, he established the link between lightning and electricity. Resulting from these experiments, he invented the lightning rod, which we still use today in various configurations.

Benjamin Franklin undersøgte elektricitet og testede teorier ved hans ekstremt farligt eksperiment for at have sin søn flyve med drage gennem en storm-truede himlen. En nøgle knyttet til dragesnoren udløst og opladet en leydnerflaske, der lagres elektrisk ladning. Således etablerede han forbindelsen mellem lyn og elektricitet. Som følge af disse eksperimenter, opfandt han den lynafleder.

En lynafleder er en anordning, som placeres på taget af en bygning for at beskytte den mod de skader, lynnedslag kan forårsage. Lynet ledes ned til en jordleder, som spreder ladningen ud i jorden. Lynaflederen blev første gang konstrueret i 1752 af Benjamin Franklin.

A lightning arrestor is a device which is placed on the roof of a building to protect it from the damage lightning can cause. The lightning is led down to a ground conductor, which spreads the charge into the ground. The lightning conductor was first constructed in 1752 by Benjamin Franklin.

Nicola Tesla is show here in his laboratory in 1899 sitting under a huge arc discharge. Tesla (seated, making notes) succeeded in generating artificial lightning in his laboratory in Colorado Springs by using coils he designed for the purpose. The experiment proved his theory that the earth was a condenser, he said. Tesla never revealed the other principles he said he discovered in this laboratory, which has long since been destroyed.

Hans Christian Ørsted, often spelled Oersted in English, lived from 14 August 1777 – 9 March 1851). By the 19th century, the Danish physicist, Hans Christian Ørsted had established that electricity and magnetism were related, and their theories were unified, He discovered that wherever charges are in motion, electric current results, and magnetism results from this electric current. The source for an electric field are these electric charges, and the magnetic field is caused by electric currents, that is, electric charges in motion.

He was a Danish physicist and chemist who <u>discovered that electric currents create magnetic fields, and was the first to establish the connection between electricity and magnetism</u>. That relationship is known today as Oersted's Law. He shaped post-Kantian philosophy and advances in science throughout the late 19th century. The Ørsted Park in Copenhagen was named after Ørsted in 1879. The streets H.C. Ørsteds Vej in Frederiksberg and H. C. Ørsteds Allé is located in Galten are also named after him.

The buildings that are home to the Department of Chemistry and the Institute for Mathematical Sciences at the University of Copenhagen's North Campus are named the H.C. Ørsted Institute, after him. A student's dormitory named H. C. Ørsted Kollegiet is in Odense. The first Danish satellite, launched 1999, was named after Ørsted.

HC Ørsted, ofte stavet Ørsted på engelsk, levede fra 14 August 1777 - marts 9, 1851). Han var en dansk fysiker og kemiker, der <u>opdagede at elektriske strømme skaber magnetiske felter, som var den første forbindelser fundet mellem elektricitet og magnetisme.</u> Han er i dag kendt for Ørsted lov. Han formede post-kantianske filosofi og fremskridt inden for videnskab igennem slutningen af det 19. århundrede. Ørsted Park i København, som er opkaldt efter Ørsted i 1879. Gaderne H. C. Ørsteds Vej på Frederiksberg og H. C. Ørsteds Allé ligger i Galten derfor er opkaldt efter ham.

Bygningerne har er hjemsted for Kemisk Institut og Institut for Matematiske Fag ved Københavns Universitets Nørre Campus er navngivet H. C. Ørsted Instituttet, efter ham. En studerendes sovesal, ved navn H. C. Ørsted Kollegiet er i Odense. Den første danske satellit, der blev lanceret i 1999, som er opkaldt efter Ørsted.

In 1824, Ørsted founded *Selskabet for Naturlærens Udbredelse* (SNU), a society to disseminate knowledge of the natural sciences. He was also the founder of predecessor organizations which eventually became the Danish Meteorological Institute and the Danish Patent and Trademark Office. Ørsted was the first modern thinker to explicitly describe and name the "thought experiment."

A leader of the so-called Danish Golden Age, Ørsted was a close friend of Hans Christian Andersen and the brother of politician and jurist Anders Sandøe Ørsted, who eventually served as Danish prime minister (1853–54).

The oersted (Oe), the cgs unit of magnetic H-field strength, is named after him.

Ørsted was born in Rudkøbing. As a young boy Ørsted developed his interest in science while working for his father, who owned a pharmacy.

Hans Christian Ørsted, född den 14 augusti 1777 i Rudkøbing på Langeland, död den 9 mars 1851 i Köpenhamn, var en dansk fysiker och kemist, bror till statsminister Anders Sandøe Ørsted och farbror till botanikern med samma namn.

He and his brother Anders received most of their early education through self-study at home, going to Copenhagen in 1793 to take entrance exams for the University of Copenhagen, where both brothers excelled academically. By 1796 Ørsted had been awarded honors for his papers in both aesthetics and physics. He earned his doctorate in 1799 for a dissertation based on the works of Kant entitled "The Architectonics of Natural Metaphysics." Home schooling really works for motivated students!

Han og hans bror Anders fik det meste af deres tidlige uddannelse gennem selvstudium hjemme, går til København i 1793 for at tage optagelsesprøver for Københavns Universitet, hvor begge brødre udmærkede fagligt. Ved 1796 Ørsted havde fået tildelt æresbevisninger for hans papirer i både æstetik og fysik. Han tog sin doktorgrad i 1799 for en afhandling baseret på værker af Kant med titlen "Arkitektoniken af Natural Metafysik".

In 1801 Ørsted received a travel scholarship and public grant which enabled him to spend three years travelling across Europe.
In Germany he met Johann Wilhelm Ritter, a physicist who believed there was a connection between electricity and magnetism. This made sense to Ørsted since he believed in **Kantian ideas about the unity of nature and that deep relationships existed between natural phenomena.**

I 1801 modtog Ørsted et rejselegat og offentlige tilskud, der gjorde ham i stand til at tilbringe tre år på at rejse i hele Europa. I Tyskland mødte han Johann Wilhelm Ritter, en fysiker, der mente, at der var en sammenhæng mellem elektricitet og magnetisme. Det gav mening for Ørsted siden han troede på kantianske ideer om enheden af natur og der forelå dybe relationer mellem fysiske fænomener.

The concept that there was unity of matter, energy, and spiritual phenomenon was widespread in the European scientific communities, and still fascinates and inspires researchers to search for meaning and order in the universe. It inspired the Danish-German scientist, Gustavus Detlef Hinrichs, to find mathematical relationships between the orbits of the planets, which were like the rotations of electrons around the nucleus of atoms. This then led to his development of one of the first models of the periodic nature of the chemical elements.

Some of the most illuminating ideas concerning spiritual and material interrelationships were developed in the 20th Century by the Dane, Martinus. He interpreted the difference between material science and the spiritual science as follows:

"There is, however, this difference between the two sciences, that while material science can be understood and accepted by intelligence and thus appear an actual fact, spiritual science can certainly not be accepted by intelligence alone, even if said mental capacity naturally constitutes a strongly contributory factor. Here, in the first place, one must have an extremely well developed potentiality for benevolence and love as well as an indispensable intuitive capacity for cosmic perception."

"The holy spirit, which thus constitutes the very highest science, cannot possibly belong to a certain nation, it cannot possibly be Danish or German, it cannot be American or English, just as it cannot be theosophical, anthroposophical, spiritualistic or Martinian. It is to the very highest degree international, indeed, interplanetary or in every imaginable way non-individual. It is the life in which we all life and move and have our being. It is the omnipresent "Spirit of God that moves upon the face of the waters.""

Martinus, Quoted from "Spiritual science" – English Kosmos no.3 – 2000

http://www.martinus-on-tour.info/1.php?id=3&page1=The%20Sciences&sprog=eng

"In the history of science, ever since the famous trial of Galileo, it has repeatedly been claimed that scientific truth cannot be reconciled with the religious interpretation of the world. Although I am now convinced that scientific truth is unassailable in its own field, I have never found it possible to dismiss the content of religious thinking as simply part of an outmoded phase in the consciousness of mankind, a part we shall have to give up from now on, Thus in the course of my life I have repeatedly been compelled to ponder on the relationship of these two regions of though, for I have never been able to doubt the reality of that to which they point."

— Werner Heisenberg

Their conversations drew Ørsted into the study of physics. He became a professor at the University of Copenhagen in 1806 and continued his research with electric currents and acoustics. Under his guidance, the University developed a comprehensive physics and chemistry program and established new laboratories.

Deres samtaler trak Ørsted til studiet af fysik. Han blev professor ved Københavns Universitet i 1806 og fortsatte sin forskning med elektriske strømme og akustik. Under hans ledelse universitetet udviklet en omfattende fysik og kemi program og etableret nye laboratorier.

In 1800, Alessandro Volta invented a galvanic battery inspiring Ørsted to think about the nature of electricity and to conduct his first electrical experiments.

I 1800, Alessandro Volta har opfundet en galvanisk batteri og inspirerende Ørsted at tænke om naturen af elektricitet og til at udføre sine første elektriske eksperimenter.

En 1800, Alessandro Volta a inventé une batterie galvanique et inspiré Ørsted à penser à la nature de l'électricité et à conduire ses premières expériences électriques.

Between 1800 and 1803, he visited to Germany, France and Holland for lectures. Ørsted welcomed William Christopher Zeise to his family home in autumn 1806; taking the then young chemist (and fellow son of a pharmacist) under his care and giving him encouragement while offering him a position as his lecturing assistant. In 1812 he again visited Germany and France after publishing a manual called *Videnskaben om Naturens Almindelige Love* and *Første Indledning til den Almindelige Naturlære* (1811).

Zwischen 1800 und 1803 besuchte er Deutschland, Frankreich und Holland um Vorträge zu hören. Ørsted begrüßte William Christopher Zeise zu seinem Familienhaus im Herbst 1806; indem er den damaligen jungen Chemiker (auch Sohn eines Apothekers) unter seine Obhut nahm und ihm Ermutigung gab, indem er Zeise eine Stelle als seinen Lehrbeauftragten anbot. Im Jahre 1812 besuchte er erneut Deutschland und Frankreich, nachdem er ein Handbuch mit dem Titel *Videnskaben om Naturens Almindelige Love und Første Indledning til den Almindelige Naturlære (1811)* veröffentlicht hatte.

In Berlin he wrote his famous essay on the identity of chemical and electrical forces in which he first stated the connection existing between magnetism and electricity. In Paris, he translated that essay into Latin with Marcel de Serres.

A Berlin, il a écrit son célèbre essai sur l'identité des forces chimiques et électriques dans lequel il a d'abord exposé la connexion existant entre le magnétisme et l'électricité. A Paris, il traduit cet essai en latin avec Marcel de Serres.

I 1800, Alessandro Volta opfundet en galvanisk batteri inspirerende

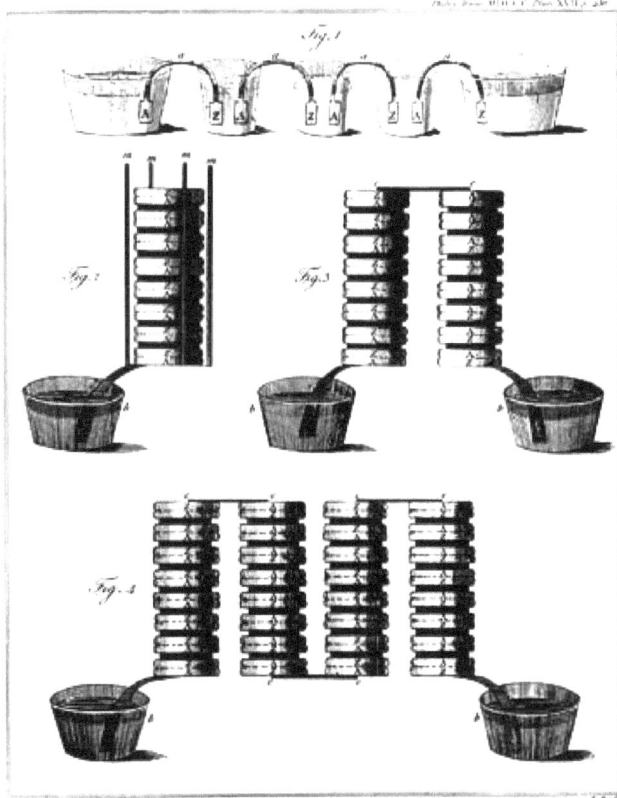

Ørsted at tænke natur elektricitet og til at udføre sine første elektriske eksperimenter. Mellem 1800 og 1803, besøgte han til Tyskland, Frankrig og Holland til foredrag. Ørsted hilste William Christopher Zeise til sin familie hjemme i efteråret 1806; tager derefter den unge kemiker (og kollegassøn af en farmaceut) under hans omsorg og give ham opmuntring samtidig tilbyde ham en stilling som hans forelæsninger assistent. I 1812 besøgte han igen Tyskland og Frankrig efter offentliggørelse af en håndbog kaldet *Videnskaben*

om Naturens Almindelige Kærlighed og *Første Indledning to the Almindelige Naturlære (1811).*

Illustration from *On the Electricity Excited by the Mere Contact of Conducting Substances of Different Kinds*, Alessandro Volta's paper announcing his invention of the wet pile in the Philosophical Transactions of the Royal Society, 1800.

I Berlin skrev han sin berømte essay om identiteten af kemiske og elektriske kræfter, hvor han først anførte forbindelsen består mellem magnetisme og elektricitet. Også, i Paris oversatte han, en essay i latin med Marcel de Serres.

The Royal Society of London gave him the Copley Medal and the French Academy awarded him with 3,000 gold francs. Ørsted was just 43 when he made this great discovery. He established the Royal Polytechnic Institute in 1829 of which he was the first director.

The Royal Society of London gav ham copleymedaljen og den franske Akademi tildelt ham med 3.000 guld francs. Ørsted var kun 43, da han gjorde denne store opdagelse. Han etablerede Royal Polytechnic Institute i 1829, hvor han var den første direktør.

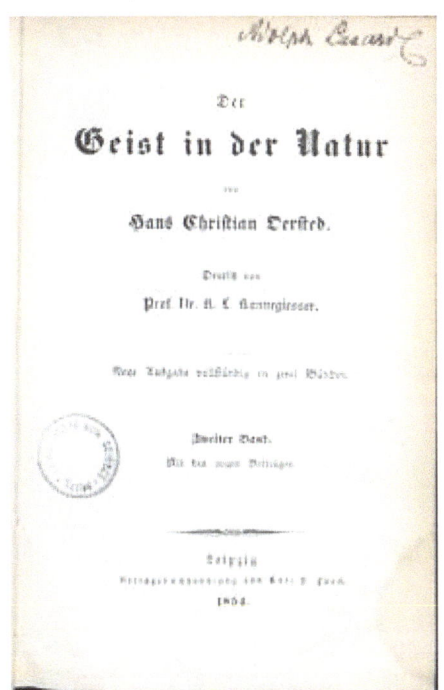

Hans Christian Ørsted, *Der Geist in der Natur*, 1854

On 21 April 1820, during a lecture, Ørsted noticed a compass needle deflected from magnetic north when an electric current from a battery was switched on and off, confirming a direct relationship between electricity and magnetism. His initial interpretation was that magnetic effects radiate from all sides of a wire carrying an electric current, as do light and heat. Three months later he began more intensive investigations and soon thereafter published his findings, showing that an electric current produces a circular magnetic field as it flows through a wire. This discovery was not due to mere chance, since Ørsted had been looking for a relation between electricity and magnetism for several years. The special symmetry of the phenomenon was possibly one of the difficulties that retarded the discovery.

Den 21. april 1820, under en forelæsning, Ørsted bemærket en kompasnål afbøjet når en elektrisk strøm fra et batteri blev tændt og slukket, bekræfter en direkte sammenhæng mellem elektricitet og magnetisme fra magnetisk nord. Hans oprindelige fortolkning var, at magnetiske virkninger udstråle fra alle sider af en ledning og transporterer en elektrisk strøm, som gør lys og varme. Tre måneder senere begyndte han mere intensive undersøgelser og snart derefter offentliggjort sine resultater, der viser, at en elektrisk strøm frembringer et cirkulært magnetfelt, som det flyder gennem en ledning. Denne opdagelse var ikke på grund af en tilfældighed, da Ørsted havde været på udkig efter en relation mellem elektricitet og magnetisme i flere år. Den særlige symmetri af

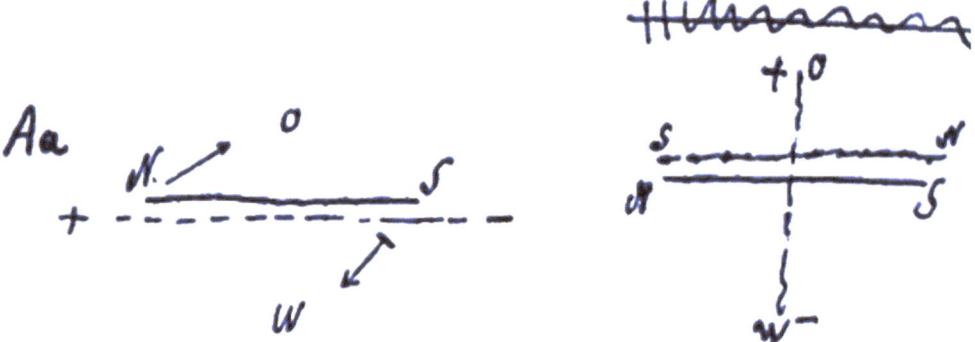

fænomenet var muligvis en af de vanskeligheder, retarderet opdagelsen.

Ørsted's original sketches of the phenomenon are shown here. His researches were published in two Italian newspapers and were largely overlooked by the scientific community.

Ørsteds originale skitser af fænomenet er vist her. Hans undersøgelser blev offentliggjort i to italienske aviser og blev stort set overset af det videnskabelige samfund.

It is sometimes claimed that Italian Gian Domenico Romagnosi was the first person who found a relationship between electricity and magnetism, about two decades before Ørsted's 1820 discovery of electromagnetism. Romagnosi's experiments showed that an electric current from a voltaic pile could deflect a magnetic needle.

Det hævdes undertiden, at italiensk Gian Domenico Romagnosi var den
første person, der fandt en sammenhæng mellem elektricitet og
magnetisme, omkring to årtier, før Ørsteds 1820 opdagelse
af elektromagnetismen. Romagnosis eksperimenter viste, at en elektrisk
strøm fra en voltasøjle kunne aflede enmagnetisk nål.

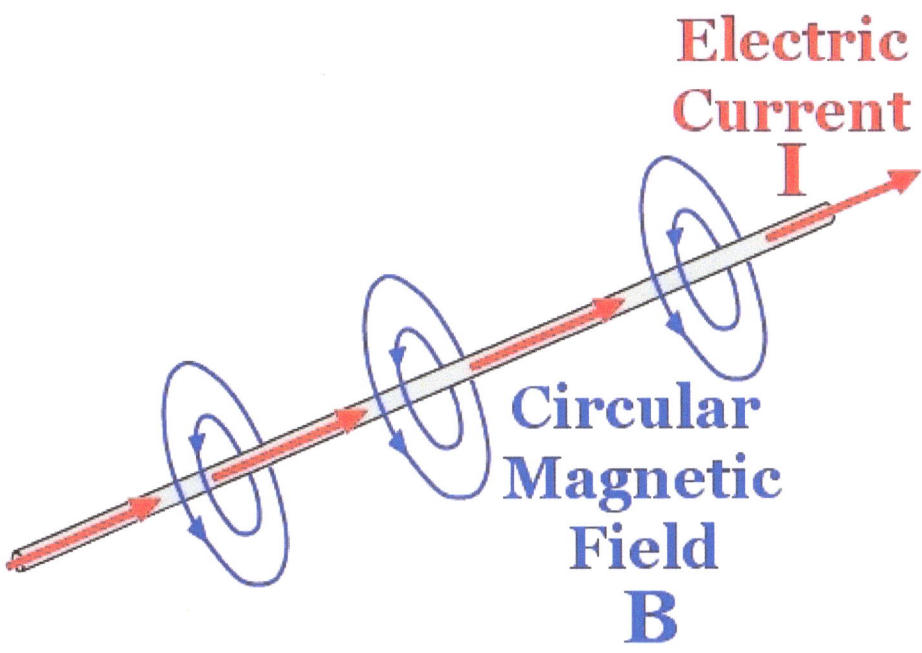

Ørsted's findings stirred much research into electrodynamics throughout
the scientific community, influencing French physicist André-Marie
Ampère's developments of a single mathematical formula to represent the
magnetic forces between current-carrying conductors. Ørsted's work also
represented a major step toward a unified concept of energy.

"Oersted *was searching* for the connection between those
two great forces of nature. His previous writings bear witness to this, and I,
who associated with him daily in the years 1818 to 1819, can state from
my own experience that the thought of discovering this still mysterious
connection constantly filled his mind."

JOHAN GEORG FORCHHAMMER, 1794 – 1865

"Ørsted ledte efter forbindelsen mellem disse to store naturkræfter. Hans
tidligere skrifter vidner om dette, og jeg, der var forbundet med ham

dagligt i årene 1818-1819, kan angive, fra min egen erfaring, at tanken om at opdage dette stadig mystiske forbindelse konstant fyldt hans sind. "

Johan Georg Forchhammer, 1794 – 1865, Chemist and Geologist from Holsten

3. THE SECOND GREAT UNIFICATION IN PHYSICS

James Clerk Maxwell (13 June 1831 – 5 November 1879) was a Scottish scientist in the field of mathematical physics. His most notable achievement was to formulate the classical theory of electromagnetic radiation, bringing together for the first time electricity, magnetism, and light as manifestations of the same phenomenon. Maxwell's equations for electromagnetism have been called the "second great unification in physics", after the first one realized by Isaac Newton.

James Clerk Maxwell (13 juni 1831-5 november 1879) var en skotsk forsker inden for matematisk fysik. Hans mest bemærkelsesværdige præstation var at formulere den klassiske teori af elektromagnetisk stråling, der samler for første gang elektricitet, magnetisme, og lys som manifestationer af det samme fænomen. Maxwells ligninger for elektromagnetisme er blevet kaldt den "anden store forening i fysik", efter den første realiseret af Isaac Newton.

James Clerk Maxwell (13. Juni 1831 - 5. November 1879) war ein schottischer Wissenschaftler auf dem Gebiet der mathematischen Physik. Sein bemerkenswerte Leistung war die Formulierung der klassischen Theorie der elektromagnetischen Strahlung, die erstmals Elektrizität, Magnetismus und Licht als Erscheinungen desselben Phänomens zusammenbrachte. Die Maxwellschen Gleichungen für den Elektromagnetismus wurden nach der ersten von Isaac Newton, als "zweite große Vereinigung in der Physik" bezeichnet.

James Clerk Maxwell (júní 13, 1831 - 5. Nóvember 1879) var skoskur vísindamaður á sviði stærðfræði og eðlisfræði. Mest áberandi velgengni hans var að móta klassíska kenningu rafsegulgeislun, uppeldi saman í fyrsta skipti rafmagn, segulmagn, og ljós og birtingarmyndir sama fyrirbæri. Jöfnur Maxwells um rafsegulfræði hafa verið kallaðir "seinni mikla sameiningu í eðlisfræði", eftir fyrsta innleystur af Isaac Newton.

Maxwell's equations are a set of partial differential equations that, together with the Lorentz force law, form the foundation of classical electromagnetism, classical optics, and electric circuits. They are the basis of all electric, optical and radio technologies such as power generation, electric motors, wireless communication, cameras, televisions, computers etc. Maxwell's equations describe how electric and magnetic fields are generated by charges, currents and changes of each other.

Maxwells ligninger er et sæt af partielle differentialligninger, der sammen med den Lorentzkraftlov, danner grundlaget for den klassiske elektromagnetisme, klassisk optik, og elektriske kredsløb. De er grundlaget for alle elektriske, optiske og radio teknologier som elproduktion, elmotorer, trådløs kommunikation, kameraer, fjernsyn, computere osv. Maxwells ligninger beskriver, hvordan elektriske og magnetiske felter genereres af elektriske ladninger, strømninger og ændringer af hinanden.

One important consequence of the equations is that fluctuating electric and magnetic fields can propagate at the speed of light. This electromagnetic radiation manifests itself in manifold ways from radio waves to light and X- or γ-rays. The equations are named after the physicist and mathematician James Clerk Maxwell, who between 1861 and 1862 published an early form of the equations, and first proposed that light is an electromagnetic phenomenon.

En vigtig konsekvens af ligningerne er, at fluktuerende elektriske og magnetiske felter kan forplante ved lysets hastighed. Denne elektromagnetisk stråling manifesterer sig i mangfoldige måder fra radiobølger til lys og X- eller γ-stråler. Ligningerne er opkaldt efter fysikeren og matematikeren James Clerk Maxwell, der mellem 1861 og 1862 offentliggjorde en tidlig form af ligninger, og først foreslog, at lys er en elektromagnetisk fænomen.

With the publication of *A Dynamical Theory of the Electromagnetic Field* in 1865, Maxwell demonstrated that electric and magnetic fields travel through space as waves moving at the speed of light. Maxwell proposed that light is an undulation in the same medium that is the cause of electric and magnetic phenomena. The unification of light and electrical phenomena led to the prediction of the existence of radio waves.

Working on the problem further, Maxwell showed that the equations predict the existence of waves of oscillating electric and magnetic fields that travel through empty space at a speed that could be predicted from simple electrical experiments; using the data available at the time, Maxwell obtained a velocity of 310,740,000 meters per second (1.0195×10^9 feet per second).

Arbejde på problemet nærmere viste Maxwell at ligningerne forudsige eksistensen af bølger af oscillerende elektriske og magnetiske felter, der rejser gennem det tomme rum med en hastighed, der kunne forudsiges ud fra simple elektriske eksperimenter. Ved hjælp af de tilgængelige data på det tidspunkt, Maxwell opnået en hastighed på 310,740,000 meter per sekund (1.0195×10^9 fod per sekund).

His famous twenty equations, in their modern form of four partial differential equations, first appeared in fully developed form in his textbook *A Treatise on Electricity and Magnetism* in 1873. Maxwell expressed electromagnetism in the algebra of quaternions and made the electromagnetic potential the centerpiece of his theory.

Hans berømte tyve ligninger, i deres moderne form af fire partielle differentialligninger, optrådte første gang i fuldt udviklede form, i sin lærebog *En afhandling om elektricitet og magnetisme* i 1873. Maxwell udtrykte elektromagnetisme i den algebra af quaternions og gjort den elektromagnetiske potentiale kernen i hans teori.

In 1881 Oliver Heaviside replaced Maxwell's electromagnetic potential field by 'force fields' as the foundation of electromagnetic theory. Heaviside reduced the complexity of Maxwell's theory down to four differential equations, now known collectively as Maxwell's Laws or Maxwell's equations. Per Heaviside, the electromagnetic potential field was arbitrary and needed to be "murdered." The use of scalar and vector potentials is now standard in the solution of Maxwell's equations.

I 1881 erstattet Oliver Heaviside Maxwells elektromagnetiske potentiale felt af "kraftfelter" som grundlaget for elektromagnetiske teori. Heaviside reduceret kompleksiteten af Maxwells teori ned til fire differentialligninger, nu kendt kollektivt som Maxwells love eller Maxwells ligninger. Per Heaviside, det elektromagnetiske potentielle felt var vilkårlig og skulle "myrdet". Brugen af skalar og vektor potentialer er nu standard i løsningen af Maxwells ligninger.

Formulation in the SI unit's convention

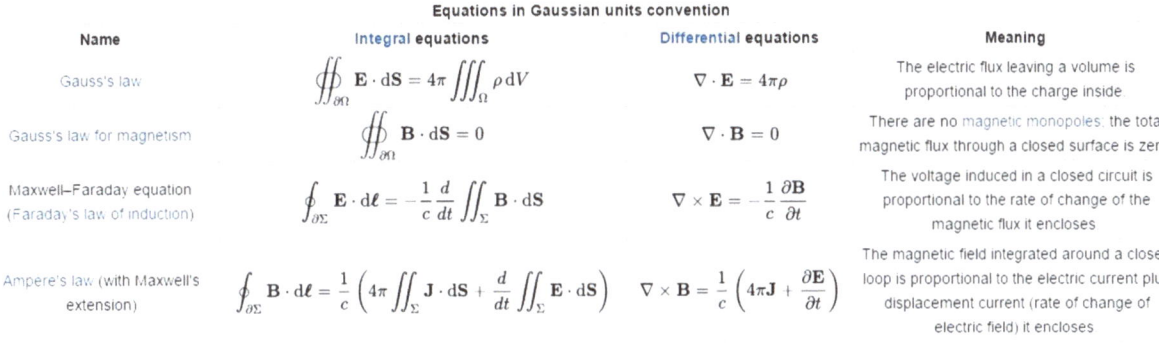

Equations in Gaussian units convention

Name	Integral equations	Differential equations	Meaning
Gauss's law	$\oiint_{\partial\Omega} \mathbf{E} \cdot d\mathbf{S} = 4\pi \iiint_\Omega \rho\, dV$	$\nabla \cdot \mathbf{E} = 4\pi\rho$	The electric flux leaving a volume is proportional to the charge inside.
Gauss's law for magnetism	$\oiint_{\partial\Omega} \mathbf{B} \cdot d\mathbf{S} = 0$	$\nabla \cdot \mathbf{B} = 0$	There are no magnetic monopoles; the total magnetic flux through a closed surface is zero.
Maxwell–Faraday equation (Faraday's law of induction)	$\oint_{\partial\Sigma} \mathbf{E} \cdot d\boldsymbol{\ell} = -\frac{1}{c}\frac{d}{dt}\iint_\Sigma \mathbf{B} \cdot d\mathbf{S}$	$\nabla \times \mathbf{E} = -\frac{1}{c}\frac{\partial \mathbf{B}}{\partial t}$	The voltage induced in a closed circuit is proportional to the rate of change of the magnetic flux it encloses
Ampere's law (with Maxwell's extension)	$\oint_{\partial\Sigma} \mathbf{B} \cdot d\boldsymbol{\ell} = \frac{1}{c}\left(4\pi \iint_\Sigma \mathbf{J} \cdot d\mathbf{S} + \frac{d}{dt}\iint_\Sigma \mathbf{E} \cdot d\mathbf{S}\right)$	$\nabla \times \mathbf{B} = \frac{1}{c}\left(4\pi\mathbf{J} + \frac{\partial \mathbf{E}}{\partial t}\right)$	The magnetic field integrated around a closed loop is proportional to the electric current plus displacement current (rate of change of electric field) it encloses

Except for Gauss's Law for Magnetism, stating that the total magnetic flux through a closed surface is zero, Maxwell's equations have accurately predicted electromagnetic phenomenon. This assumption was challenged years later with the discovery of "magnetic monopoles."

The net magnetic flux out of any closed surface is zero. This amounts to a statement about the sources of magnetic field. For a magnetic dipole, any closed surface the magnetic flux directed inward toward the south pole will equal the flux outward from the north pole. The net flux will always be zero for dipole sources.

$$\text{Magnetic flux} = \Phi = B\,A$$

Magnetic field

Area perpendicular to magnetic field B

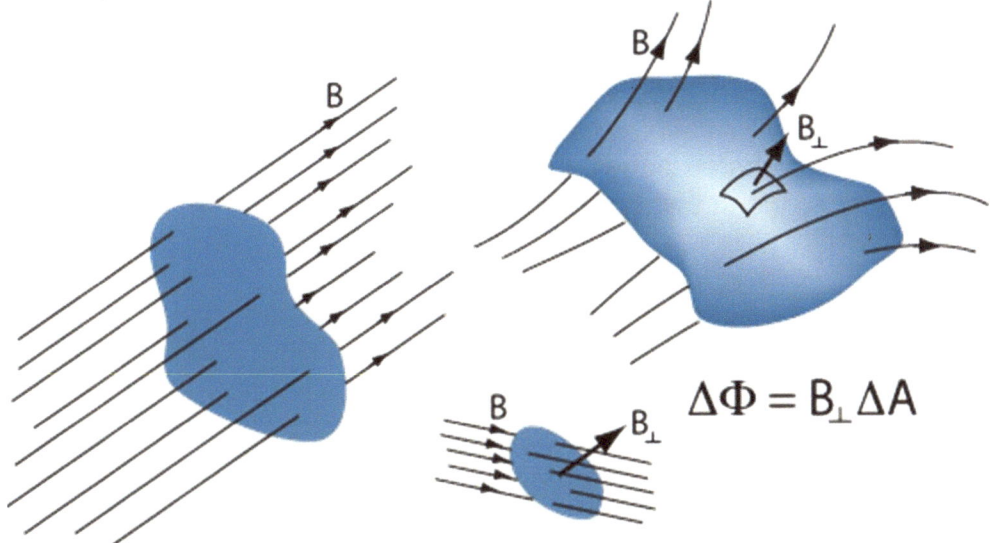

$$\Delta\Phi = B_\perp \Delta A$$

If there were a magnetic monopole source, this would give a non-zero area integral. The divergence of a vector field is proportional to the point source density, so the form of Gauss' law for magnetic fields states that there are no magnetic monopoles, which contradicts the discoveries of magnetic dipoles in the 20th Century.

4. GAUSS' LAW FOR ELECTRICITY

The electric flux out of any closed surface is proportional to the total charge enclosed within the surface. The integral form of Gauss' Law finds application in calculating electric fields around charged objects. In applying Gauss' law to the electric field of a point charge, one can show that it is consistent with Coulomb's law. While the area integral of the electric field gives a measure of the net charge enclosed, the divergence of the electric field gives a measure of the density of sources. It also has implications for the conservation of charge.

https://in.answers.yahoo.com/question/index;_ylt=AwrBT8slt4tY4T0ABhBXNyoA;_ylu=X3 oDMTEyMXI3MHF0BGNvbG8DYmYxBHBvcwMxBHZ0aWQDQjI4MjNfMQRzZWMDc2M -?qid=20080807101103AAK5VTL

In other words, the contribution to magnetic flux for a given area is equal to the area times the component of magnetic field perpendicular to the area. For a closed surface, the sum of magnetic flux is always equal to zero, if only magnetic dipoles exist, according to Gauss' law for magnetism. No matter how small the volume, the magnetic sources are always dipole sources (like miniature bar magnets), so that there are as many magnetic field lines coming in (to the south pole) as out (from the north pole).

http://hyperphysics.phy-astr.gsu.edu/hbase/magnetic/fluxmg.html

Nearly 85 years after pioneering theoretical physicist Paul Dirac predicted the possibility of the existence of magnetic monopoles, an international collaboration led by Amherst College Physics Professor David S. Hall '91 and Aalto University (Finland) Academy Research Fellow Mikko Möttönen has created, identified and photographed synthetic magnetic monopoles in Hall's laboratory on the Amherst campus. The groundbreaking accomplishment paves the way for the detection of the particles in nature, which would be a revolutionary development comparable to the discovery of the electron. This was reported in a paper co-authored by Hall, Möttönen, Amherst postdoctoral research associate Michael Ray, Saugat Kandel and Finnish graduate student Emmi Ruokokski in the journal *Nature*.

"The creation of a synthetic magnetic monopole should provide us with unprecedented insight into aspects of the natural magnetic monopole—if indeed it exists," said Hall, explaining the implications of his work.

Ray, the paper's lead author and first to discover the monopoles in the laboratory, stated: "This is an incredible discovery. To be able to confirm the work of one of the most famous physicists is probably a once-in-a-lifetime opportunity. I am proud and honored to have been part of this great collaborative effort."

Ordinarily, magnetic poles come in pairs: they have both a north pole and a south pole. As the name suggests, however, a magnetic monopole is a magnetic particle possessing only a single, isolated pole—a north pole without a south pole, or vice versa. In 1931, Dirac published a paper that explored the nature of these monopoles in the context of quantum mechanics. Despite extensive experimental searches since then, in everything from lunar samples—moon rock—to ancient fossilized minerals, no observation of a naturally-occurring magnetic monopole has yet been confirmed.

The Large Hadron Collider at CERN, the European Organization for Nuclear Research, should continue searching for magnetic monopoles, as a result of the discovery of the synthetic monopole by Hall's team. Magnetic monopoles were predicted to be quite common to early theoretical models of the Big Bang theory. Currently, a special model for the expansion of the universe in vogue, predicts that these particles will be extremely rare.

More information: Observation of Dirac Monopoles in a Synthetic Magnetic Field, M. W. Ray, E. Ruokokoski, S. Kandel, M. Möttönen, and D. S. Hall, Nature, 2014: dx.doi.org/10.1038/nature12954

Also, read: https://phys.org/news/2014-01-physicists-synthetic-magnetic-monopole-years.html#jCp

The Earth has a magnetosphere that affects the life of most creatures on Earth. Earth's magnetism is very weak, from 0.3 gauss at the Equator to 0.7 gauss at the Poles. It has been well known that many living organisms are sensitive to the magnetic field of the earth, including some geomagnetically sensitive bacteria. The migration of Monarch butterflies, Canadian geese, and salmon to their breeding grounds is evidence of their response to the earth's magnetic field. Stray and lost dogs and cats have been known to travel hundreds of miles to their home locations using geomagnetic sensitivities. Studying how bacteria interact with geomagnetic fields has contributed to understanding how higher organisms react also. Geomagnetically responsive animals have something in common with the ferromagnetic detection systems in bacteria. Cell motility caused by a magnetic field is called "magnetotaxis." "Taxis" refers to the effect of the magnetic field to influence the swimming direction, but not the absolute velocity of cell movement. When a magnet was brought near a microscope, the cells' motion was demonstrably affected. Hundreds of swimming cells were instantly turned around and repelled from the end of the magnet.

http://www.roaringlionpublishing.com/tony_uploads/Magnetotactic_Bacteria.pdf

A magnetotactic green alga belonging to the genus Chlamydomohas was discovered by H. Lins de Barros, et al, in a polluted brackish coastal lagoon of Brazil. These eukaryotic cells possess a "permanent magnetic dipole moment estimated to be ten times greater than those of magnetotactic bacterial cells." Because of their cell polarity, they are steered downward in the geomagnetic field.

https://phys.org/news/2014-01-physicists-synthetic-magnetic-monopole-years.html

Researchers discovered magnetic bacteria living in the ponds and lakes, presenting a chain of magnetic crystals inside their cells. Those located in the Northern Hemisphere swim in the direction of the magnetic north, while those from the Southern Hemisphere swim in the direction of the magnetic south (these bacteria live in environments with poor oxygen supply) … In 2004, tiny deposits of a mineral called magnetite (lodestone) were discovered in the beaks of pigeons and bobolink (a North American songbird). If a small magnet impairs their reception of the Earth's magnetism or the area has a natural disturbance of the terrestrial magnetism, pigeons get disoriented and cannot find the way back on long distances.

European robins and 20 songbirds and pigeons were proven to rely on Earth's magnetic field to navigate during their seasonal migrations. In fact, migratory birds are believed to use magnetism, rather than sight, to make their long journeys. Birds and many non-mammal magnetic sensitive animals are known to perceive the inclination, **using the Earth's magnetic field inclination to assess relative latitude.** This angle is 90 degrees at the poles (perpendicular to the Earth) and 0 degrees in the equatorial area (parallel to the Earth). There are bird species, like the Arctic Tern, which use this information to make annual journeys from the North Pole to the South Pole and back.

The researchers put the magnetism detection on light exposure, magnetite receptors, or both. Chickens detect the terrestrial magnetism as well, and their magnetic sensors could be in the eye. This presumption is based on an experiment that revealed that the birds could orient themselves under blue light, but totally lost the sense of direction under longer-wavelength lighting.

http://news.softpedia.com/news/Animals-and-the-Magnetic-Field-of-the-Earth-81528.shtml

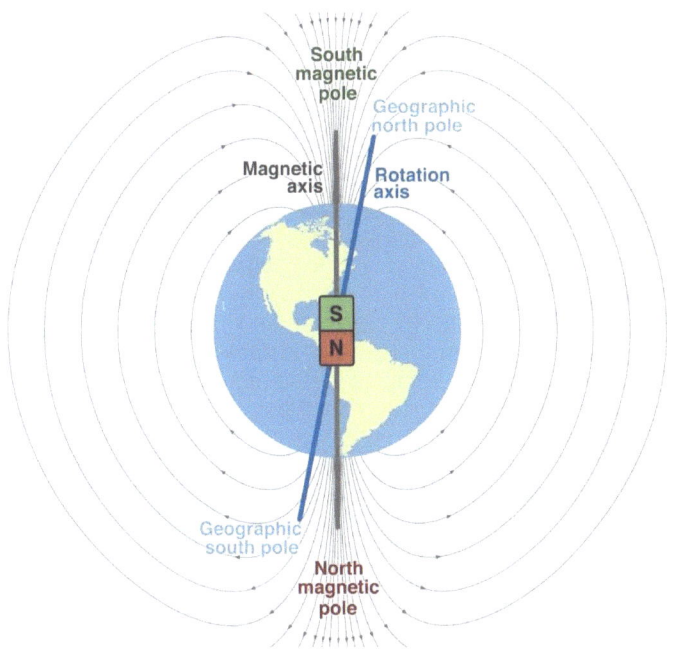

Entangled quantum state of magnetic dipoles

"Free magnetic moments usually manifest themselves in Curie laws, where weak external magnetic fields produce magnetizations that vary as the reciprocal of the temperature (1/T). For a variety of materials that do not display static magnetism, including doped semiconductors and certain rare-earth intermetallics, the 1/T law is replaced by a power law T- with < 1. Here we show that a much simpler material system—namely, the insulating magnetic salt $LiHo_xY_{1-x}F_4$—can also display such a power law. Moreover, by comparing the results of numerical simulations of this system with susceptibility and specific-heat data, we show that both energy-level splitting and quantum entanglement are crucial to describing its behaviour. The second of these quantum mechanical effects—entanglement, where the wavefunction of a system with several degrees of freedom cannot be written as a product of wavefunctions for each degree of freedom—becomes visible for remarkably small tunnelling terms, and is activated well before tunnelling has visible effects on the spectrum. This finding is significant because it shows that entanglement, rather than energy-level redistribution, can underlie the magnetic behaviour of a simple insulating quantum spin system."

Ghosh, S.; Rosenbaum, T.F.; Aeppli, G.; Coppersmith, S.N.; (2003) Entangled quantum state of magnetic dipoles. Nature, 425 (6953) pp. 48-51. 10.1038/nature01888.

To find a magnetic monopole is a Holy Grail of physics. A magnetic monopole is the magnetic version of a charged particle like an electron, and for the last 70 years physicists have believed that one might exist somewhere in the universe.

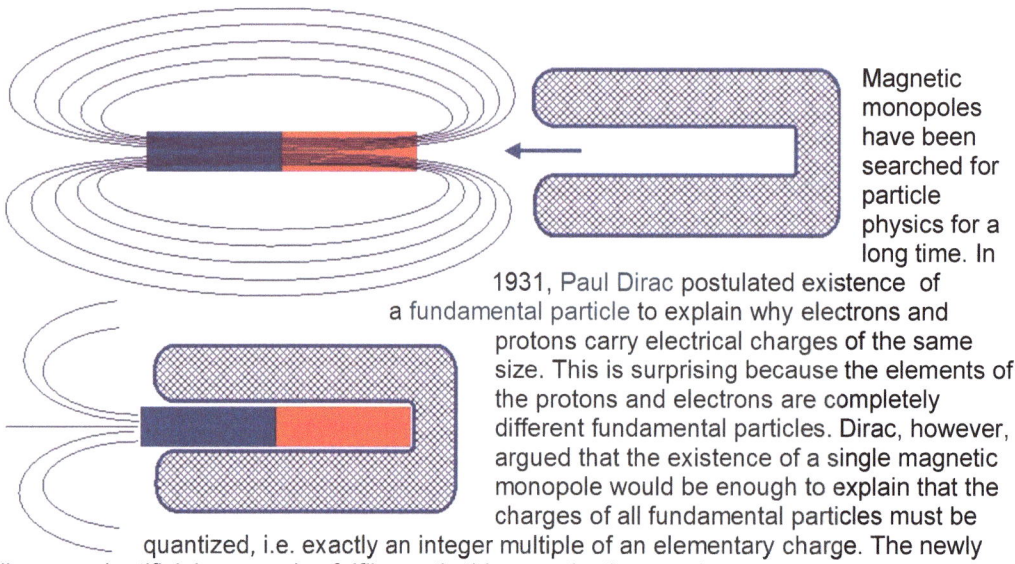

Magnetic monopoles have been searched for particle physics for a long time. In 1931, Paul Dirac postulated existence of a fundamental particle to explain why electrons and protons carry electrical charges of the same size. This is surprising because the elements of the protons and electrons are completely different fundamental particles. Dirac, however, argued that the existence of a single magnetic monopole would be enough to explain that the charges of all fundamental particles must be quantized, i.e. exactly an integer multiple of an elementary charge. The newly discovered artificial monopoles fulfil exactly this quantization requirement.

Read more at: https://phys.org/news/2013-05-artificial-magnetic-monopoles.html#jCp

The monopoles discovered are not that Holy Grail, but are the next best thing. Rather than existing throughout the universe, they only exist within a special type of material called `spin ice'. They can be imagined as the north and south poles of magnets, but free to float around independently within the material. However, someone living within in a block of spin ice would think that these are exactly those magnetic monopoles long sought by physicists…. The research shows how certain real materials, in this case spin ice, create within themselves things that resemble the basic particles by which the universe is composed. "The amazing thing about spin ice monopoles" Prof. Bramwell states, "is how perfect they are: they really do look just like those monopoles expected to exist somewhere in the universe. Why nature should reproduce a mini-universe within a material, we do not yet know" …

As well as having implications for fundamental physics, the monopoles could be harnessed in the way that electrical charges are, for technology. Scientists are currently researching the question of whether magnetic whirls could be used in the production of computer components. If possible, they would also have to create and destroy whirls: magnetic monopoles would then play an important role in this.

Samt at have konsekvenser for grundlæggende fysik, kunne de monopoler udnyttes på den måde elektriske ladninger gjorde for teknologi. Forskere i øjeblikket forsker i spørgsmålet om, hvorvidt magnetiske hvirvler kunne bruges i produktionen af computerens komponenter. Hvis det er muligt, vil de derfor nødt til at skabe og ødelægge hvirvler: magnetiske monopoler ville så spille i vigtig rolle i denne.

Read more at: https://phys.org/news/2013-05-artificial-magnetic-monopoles.html#jCp "A magnetic version of electricity is a long way off" says Prof. Bramwell, "but these results are an important first step".

https://www.london-nano.com/research-and-facilities/highlight/magnetic-monopoles-discovered-by-lcn-scientists

The discovery of magnetic monopoles changed all that! Then along came Dr. Konstatin Meyl, and the revelation of "The Physical Basis of Life."

Meyl wurde 1984 an der Universität Stuttgart promoviert und wurde 1986 als Professor an die Fachhochschule Furtwangen berufen, die inzwischen in Hochschule Furtwangen umbenannt wurde. Dort ist er Dozent für Leistungselektronik sowie der Antriebs- und Steuerungstechnik.

Meyl was awarded his doctorate at the University of Stuttgart in 1984 and was appointed professor at the Fachhochschule Furtwangen in 1986, which was renamed Furtwangen University of Applied Sciences. There he is a lecturer for power electronics as well as drive train technology and control engineering.

Meyl var tildelt sin doktorafhandling ved universitetet i Stuttgart i 1984, og der var udnævnt professor ved Fachhochschule Furtwangen i 1986, hvilket hvad omdøbt Furtwangen University of Applied Sciences. Der er han foredragsholder for præstationselektronik samt gearsystemsteknologi og reguleringsteknik.

Daneben wirbt Meyl für seine Überzeugung, dass die klassische Elektrodynamik auf der Grundlage der Maxwellschen Gleichungen unvollständig sei und durch eine Theorie ersetzt werden müsse, die er selbst begründet hat. Zentrale Begriffe darin sind „Potentialwirbel", „Skalarwelle" oder „Neutrinopower" (auch in Bezug auf die Expansionstheorie der Erde). Er behauptet, damit eine einheitliche Feldtheorie entwickelt zu haben, aus der alle bekannten Wechselwirkungen ableitbar sind. **Sie steht allerdings im Widerspruch zu den etablierten physikalischen Theorien der Elektrodynamik, widerspricht auch der Einstein'schen speziellen Relativitätstheorie.**

Meyl also argues that classical electrodynamics is incomplete, based on Maxwell's equations and must be replaced by a theory which he himself has established. The main terms are "potential vertebra", "scalar wave" or "neutrinopower" (regarding the theory of expansion of the earth). He claims to have developed a uniform field theory from which all known interactions are derivable. However, it is in contradiction to the established physical theories of electrodynamics, which also contradicts Einstein's special theory of relativity.

The orthodox scientific community does not uniformly accept Meyl's theory and demonstrated evidence of the presence of scalar waves, but then Galileo and Tesla were not acknowledged in their day either!

Die Hochschule Furtwangen hat sich von Meyls Ideen distanziert, darauf hingewiesen, dass sie „wissenschaftlich und methodisch nicht anerkannt, sondern in der Fachwelt äußerst umstritten sind" und klargestellt, dass sie nicht Gegenstand seiner Tätigkeiten an der Fakultät sind. Weiterhin darf Meyl über seine Theorien keine Vorlesungen an der Hochschule halten. Kritiker Meyls sind der Mathematikprofessor Gerhard Bruhn und Klaus Keck. Bruhn hat nach eigenen Angaben mathematische Fehler und Widersprüche aufgezeigt, die Meyl bei der Herleitung seiner Theorie unterlaufen seien. Außerhalb der Skeptikerbewegung hat sich Thomas Eibert für die Forschungsgemeinschaft Funk mit einigen Arbeiten Meyls auseinandergesetzt.

The University of Furtwangen has distanced itself from Meyl's ideas, pointing out that they are "not scientifically and methodically not recognized but highly controversial in the professional world" and clarified that they are not the subject of his activities at the faculty. Furthermore, Meyl is not permitted to give lectures at the university on his theories. Meyl's critics include the mathematics professor Gerhard Bruhn and Klaus Keck. Bruhn has shown mathematical errors and contradictions made by Meyl in his derivation of his theory. Beyond these sceptics, Thomas Eibert has dealt with some of Meyl's works for the Forschungsgemeinschaft Funk.

Das *Institut für Gravitationsforschung* der Göde-Wissenschaftsstiftung hat den von Meyl vertriebenen Experimentierbausatz untersucht und kam zu dem Ergebnis, dass sich alle Beobachtungen im Rahmen der klassischen Elektrodynamik durch Übertragung mit transversalen elektromagnetischen Wellen erklären ließen.

The Institute for Gravitational Research at the Göde Science Foundation has investigated Meyl's experimental setup and concluded that all observations in classical electrodynamics could be explained by transmission with transversal electromagnetic waves. Clearly a controversy exists which future scientists must resolve!

Konstantin Meyl verändert die Expansionstheorie der Erde, die im wissenschaftlich-akademischen Umfeld nicht mehr vertreten wird, dahingehend, dass nicht Äther, sondern die Absorption von Neutrinos aus dem Weltall die Erde expandieren ließen.

Konstantin Meyl changes the theory of the expansion of the earth, which is no longer represented in the scientific-academic environment, in the sense that it is not ether but the absorption of neutrinos from the universe that allow the earth to expand.

Isaac Newton formulerte lova om gravitasjon i *Philosophiae Naturalis Principia Mathematica* (publisert 5. juli 1687). Ho seier at gravitasjonskrafta som verkar mellom to lekamar, er proporsjonal med produktet av massen deira, og omvendt proporsjonal med kvadratet av avstanden mellom dei.

Isaac Newton formulated the law of gravity in *Philosophiae Naturalis Principia Mathematica* (published 5 July 1687). It states that the gravitational force exerted between two bodies, is proportional to the product of their masses, and inversely proportional to the square of the distance between them.

$$F_1 = F_2 = G\frac{m_1 \times m_2}{r^2}$$

В рамках классической механики гравитационное притяжение описывается законом всемирного тяготения Ньютона, который гласит, что сила гравитационного притяжения между двумя материальными точками массы

$$m_1 \quad и \quad m_2 ,$$

разделёнными расстоянием r , пропорциональна обеим массам и обратно пропорциональна квадрату расстояния — то есть:
Здесь — гравитационная постоянная, равная примерно $6{,}67 \times 10^{-11}$ м³/(кг·с²).

In the framework of classical mechanics, the gravitational attraction is described by Newton's law of universal gravitation, which states that the force of gravitational attraction between two masses of material objects

m1 and m2,

separated by a distance proportional to both masses and inversely proportional to the square of the distance - that is:

Here - the gravitational constant, γ is equal to approximately 6.67×10^{-11} m³ / (kg · s²).

Lova kan skrivast slik (The Law can be written as follows):

$$G = \gamma\, m_1 m_2 / r^2$$

der (where):

- *G* er storleiken på gravitasjonskrafta mellom to objekt (G is the size of the gravitational force between two objects).

- m_1 er massen til det første objektet (hovudobjektet) (m1 is the mass of the first object (main object)).

- m_2 er massen til det andre objektet (m2 is the mass of the second object).

- r er avstanden mellom objekta (r is the distance between the objects).

- γ er gravitasjonskonstanten (γ is the gravitational constant).

Gravitationen (av latin gravis = tung) eller tyngdkraften (Gravity is from the Latin *gravis* meaning heavy, Swedish).

Meyls Artikel zur interzellulären Kommunikation mit Hilfe von Skalarwellen wurde aus zwei wissenschaftlichen Zeitschriften und einem Tagungsband entfernt, nachdem die Herausgeber festgestellt hatten, dass Meyl den Artikel mit nahezu identischem Inhalt viermal veröffentlicht hatte.

Meyl's article on intercellular communication using scalar waves was removed from two scientific journals and a conference volume after the editors had determined that Meyl had published the article with nearly the same content four times.

Furtwangens Universitet har taget afstand fra Meyl ideer, påpegede at de er "ikke videnskabeligt og metodisk anerkendt, men yderst kontroversielt i den professionelle verden" og afklaret at de ikke er omfattet af hans aktiviteter på fakultetet. Yderligere mere, Meyl har ikke lov til at holde foredrag på universitetet på hans teorier. Meyls kritikere er matematik professor Gerhard Bruhn, der er medlem af GWUP, og Klaus Keck. Bruhn har ved sine egne data, påpegede matematiske fejl og modsigelser Meyl havde lavet i afledning af hans teori. Uden for skeptikerer, har Thomas Eibert behandlet nogle af Meyls arbejdet for Research Association for Radio.

Contrary to the equations expressing Gauss's Law for Magnetism, Konstatin Meyl has shown that magnetic monopoles do exist! This has profound implications for explaining the "physical basis of life", from an electromagnetic, as well as a biological basis. Potential vortexes are an essential component of a scalar waves, as discovered in 1990. The basic approach for an extended field theory was confirmed in 2009 with the discovery of magnetic monopoles. For the first time, this provides the opportunity to explain the physical basis of life not only from the biological discipline. Nature covers the whole spectrum of known scientific fields of research, and interdisciplinary understanding is required to explain its complex relationships.

I motsats till de ekvationer som uttrycker den Gauss Lag för Magnetism, har Konstatin Meyl visat att magnetisk monopoler existerar! Detta har djupgående implikationer för att förklara den "fysiska grunden för livet", från en elektromagnetisk , samt en biologisk grund. Potentiella virvlarna är en viktig del av en skalär vågor, som upptäcktes 1990. Den grundläggande tillvägagångssättet för en förlängd field teori war bekräftade i 2009 med upptäckten av magnetisk monopol. För första gången ger detta möjlighet att förklara den fysiska grunden för livet, inte bara från den biologiska disciplin. Natur täcker hela spektrat av kända vetenskapligt fields av forskning och tvärvetenskaplig förståelse krävs för att förklara sina komplexa relationer.

Im Gegensatz zu den Gleichungen, die das Gaußsche Gesetz für Magnetismus ausdrücken, hat Konstatin Meyl gezeigt, daß magnetische Monopole existieren! Dies hat tiefgreifende Konsequenzen für die Erläuterung der "physischen Basis des Lebens", sowohl auf elektromagnetischer als auch auf biologischer Basis. Potentielle Wirbel sind eine wesentliche Komponente einer Skalarwellen, wie sie 1990 entdeckt wurde. Der grundlegende Ansatz für eine erweiterte Feldtheorie wurde 2009 mit der Entdeckung magnetischer Monopole bestätigt. Damit ist erstmals die Möglichkeit geschaffen, die physische Basis des Lebens nicht nur aus der biologischen Disziplin zu erklären. Die Natur deckt das gesamte Spektrum der bekannten Wissenschaftsforschungsbereiche ab, und es bedarf interdisziplinäres Verständnis, um ihre komplexen Zusammenhänge zu erläutern.

I modsætning til de ligninger, der udtrykker den Gauss lov for magnetisme, har Konstatin Meyl vist, at magnetiske monopoler faktisk eksisterer. Dette har stor betydning for at forklare den "fysiske grundlag af liv", fra et elektromagnetisk, samt en biologisk basis. Potentielle hvirvler er en væsentlig bestanddel af en skalar bølger, som blev opdaget i 1990. Den grundlæggende tilgang til en udvidet felt teori blev bekræftet, i 2009 med opdagelsen af magnetiske monopoler. For første gang, dette giver mulighed for at forklare den fysiske grundlag for livet, ikke kun fra den biologiske disciplin. Naturen dækker hele spektret af kendte videnskabelige forskningsfelter, og tværfaglig forståelse er nødvendig for at forklare sine komplekse sammenhænge.

To get to the bottom of the theories advocated and developed by Meyl, it is useful to follow the development of electromagnet theory from Oersted, to Faraday and Heaviside, greatly illuminated by the revolutionary experiments and patents of Nikola Tesla, and culminating with the scientific work of Meyl and Bearden. The applications of electromagnetic-gravitational and scalar wave technologies will be examined, particularly with respect to the weapons of mass destruction developed by the Soviets during the Cold War. Additionally, some biomedical applications of scalar technology which can have helpful benefits to humankind, will also be discussed.

5. FARADAY AND THE CONSTRUCTION OF THE ELECTRIC GENERATOR

Michael Faraday began the first practical application of electromagnet theory with his construction of the electric generator. The horseshoe-shaped magnet (A) created a magnetic field through the disk (D). When the disk was turned, this induced an electric current radially outward from the center toward the rim. The current flowed out through the sliding spring contact m, through the external circuit, and back into the center of the disk through the axle.

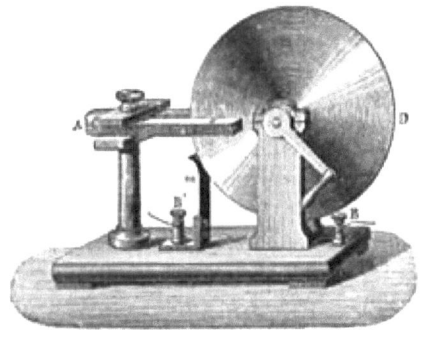

Faradays skive var af den første elektriske generator. Det hesteskoformede magnet (A) skabte et magnetfelt gennem skiven (D). Når disken blev vendt, dette inducerede en elektrisk strøm radialt udad fra midten og ud mod kanten. Den nuværende strømmede ud gennem glidende kontaktfjederen m, gennem det ydre kredsløb, og tilbage i midten af skiven gennem akslen.

Den første brugbare generator blev konstrueret 1867 i Tyskland, ved konstruktionen udnyttedes H.C. Ørsteds opdagelse af elektromagnetismen.

The first usable generator was constructed in 1867 in Germany, and the design exploited the Hans Christian Oersted's discovery of electromagnetism.

Werner von Siemens, der bereits als Industrieller erfolgreich gewesen war, kümmerte sich nun mit Nachdruck um die Weiterentwicklung der elektrischen Energietechnik. So nutze er für die Verbesserungen bereits elektrisch betriebener Maschinen, die jedoch eine externe Stromquelle benötigten, damit sie Strom liefern konnten. 1867 stellte Werner von Siemens sein Verbessertes Modell vor. Damit ermöglichte er die kostengünstige und flexible Erzeugung von Strom und begründete er die Starkstromtechnik.

Werner von Siemens, who had already been successful as an industrialist, was now taking an active part in the further development of electrical energy technology. Thus, he used the improvements of already electrically operated machines, which however needed an external current source, so they could supply electricity. In 1867, Werner von Siemens introduced his improved model. This enabled the cost-effective and flexible generation of electricity and justified the use of large power technology.

Elektrische Generatoren arbeiten mit einem Magnetfeld. Anfangs nutzte man dafür Eisenmagnete, die jedoch aufgrund ihrer Größe und ihr Gewicht der großtechnischen Elektrizitätserzeugung im Weg standen. Von Siemens entwickelte eine kleine elektromagnetische Maschine ohne Batterie und permanente Magnete, die sich in einer Richtung ohne allen Kraftaufwand in jeder Geschwindigkeit drehen ließ. Dabei wurde so ein starker elektrischer Strom erzeugt, dass deren Drahtwindungen sich schnell erhitzen.

Electric generators operate with a magnetic field. Initially, they used iron magnets, which were, however, because of their size and weight of large-scale electricity production. Siemens developed a small electromagnetic machine without a battery and permanent magnets, which could be turned in one direction without any effort at any speed. A strong electric current was generated in such a way that their wire windings quickly heated.

One of the first 3-phase generators ever built 1887: Built by August Haselwander

Location: Deutsches Museum Munich, DE

U.S. Nuclear Regulatory Commission image of a modern steam turbine generator (STG).

In electricity generation, a **generator** is a device that converts mechanical energy to electrical energy for use in an external circuit. Sources of mechanical energy include steam turbines, gas turbines, water turbines, internal  combustion engines and even hand cranks. The first electromagnetic generator, the Faraday disk, was built in 1831 by British scientist Michael Faraday. Today, generators provide nearly all the power for electric power grids worldwide.

Bei der Stromerzeugung ist ein Generator eine Vorrichtung, die mechanische Energie in elektrische Energie umwandelt, um sie in einer externen Schaltung zu verwenden. Zu den mechanischen Energiequellen gehören Dampfturbinen, Gasturbinen, Wasserturbinen, Verbrennungsmotoren und sogar Handkurbel. Der erste elektromagnetische Generator, die Faraday-Scheibe, wurde 1831 vom britischen Wissenschaftler Michael Faraday gebaut. Heute liefern Stromerzeuger nahezu die gesamte Stromversorgung.

I elproduktion, en generator er en anordning, der konverterer mekanisk energi til elektrisk energi til brug i et ydre kredsløb. Kilder til mekanisk energi omfatter dampturbiner, gasturbiner, vandturbiner, forbrændingsmotorer og endda hånd håndsving. Den første elektromagnetiske generator Faraday disk, blev bygget i 1831 af den britiske videnskabsmand Michael Faraday. I dag, generatorer giver næsten al energi til elektriske elnetet.

Note: Det kinesiske elnet er drevet af kulkraft, og det gør elbiler mere forurenende end tilsvarende benzinbiler. Ifølge dansk ekspert handler det om, at kineserne lige nu prioriterer uafhængighed af andres olieeksport frem for renere luft. (The Chinese power grid is powered by coal, which makes electric vehicles more polluting than comparable gasoline cars. Per Danish experts, the Chinese value the independence from foreign oil imports higher than cleaner air). Application of the energy principles discovered by Tesla and Meyl could go a long way in meeting the world energy requirements, without resorting to the continual conflict over Middle East oil resources which has resulted in the deaths of millions of people, and the squandering of trillions of dollars!

I elproduktion, är en generator en anordning som omvandlar mekanisk energi två elektrisk energi för användning i en extern krets. Källor till mekanisk energi innefattar ångturbiner, gasturbiner, vattenturbiner , förbränningsmotorer och även handvevar . Den första elektromagnetiska generatorn, den Faraday skiva, byggdes 1831 av brittisk vetenskapsman Michael Faraday. Idag, generatorer ger nästan all energi till elektriska kraftnätet.

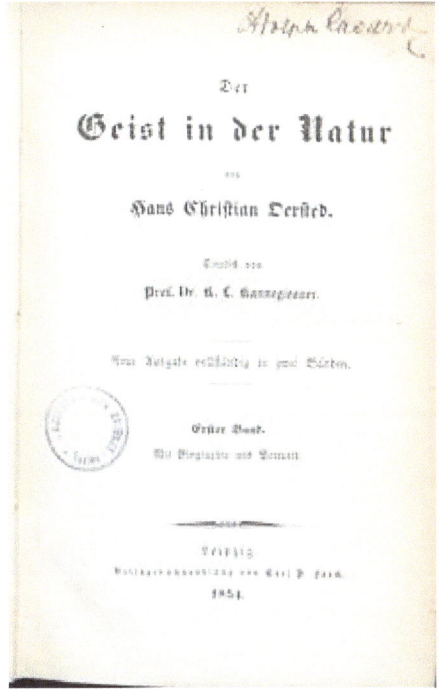

In 1825, Ørsted also made a significant contribution to chemistry by producing aluminium for the first time. While an aluminium-iron alloy had previously been developed by British scientist and inventor Humphry Davy, Ørsted was the first to isolate the element via a reduction of aluminium chloride.

In 1829, Ørsted founded Den Polytekniske Læreanstalt ('College of Advanced Technology') which was later renamed the Technical University of Denmark (DTU). Ørsted died at Copenhagen in 1851, aged 73, and was buried in the Assistens Cemetery in the same city.

The centimetre-gram-second system (CGS) unit of magnetic induction (oersted) is named for Ørsted's contributions to the field of electromagnetism.

A statue of Hans Christian Ørsted was installed in the Ørsted Park in 1880. A commemorative plaque is located above the gate on the building in Studiestræde where he lived and worked. The 100 danske kroner note issued from 1950 to 1970 carried an engraving of Ørsted.

Ørsted was also a published writer and poet. His poetry series *Luftskibet* ("The Airship") was inspired by the balloon flights of fellow physicist and stage magician Étienne-Gaspard Robert. Shortly before his death, he submitted a collection of articles for publication under the title "The Soul in Nature". The book presents Ørsted's life philosophy and views on a wide variety of issues.

Ørsted confirmed Boyles and Mariottes studies of the compressibility of air. He is generally attributed to the discovery of electromagnetism, which happened in 1820. The relationship between electricity and magnetism was already demonstrated in 1802 by Romagnosi, but this was largely unknown in scientific circles. Ørsteds discovery, however, was immediately recognized in the scientific community, and others worked on with the discovery, including Faraday, who discovered induction, while Maxwell in 1873 formulated a conclusive theory of electromagnetism.

Han konstruerede et piezometer og opdagede grundstoffet aluminium. Han var også interesseret i sprog og skabte flere nye danske ord, bl.a. ilt og brint.

He constructed a piezometer and discovered the element aluminum. He was also interested in language and created several new Danish words, including oxygen and hydrogen.

Han tilbragte sin barndom og første ungdom i Rudkøbing, indtil han sammen med sin et år yngre bror Anders i 1793 flyttede til København. Den undervisning, brødrene fik i hjemmet, var særpræget. En tysk parykmager gav dem deres første undervisning. Mens parykmageren forsøgte at indgyde dem beundring for alt hvad der var tysk, søgte en anden lærer, en norsk student, at fremhæve Norge frem for Danmark i deres øjne. Men gennem deres ualmindeligt store videbegær, erstattede de ved udbredt selvstudie i forskellige retninger, hvad der manglede i traditionel undervisning. Hvor godt de fulgte med tidens politiske og religiøse bevægelser, skildrer den yngre bror levende i *Af mit Livs og min Tids Historie*. 1794 blev de begge studenter, hvorefter de efter en kortvarig forberedelse bestod anden eksamen ved universitetet med udmærkelse i alle fag, og på grund af deres slægtskab med stifteren fik plads på Elers Kollegium.

He spent his childhood and early youth in Rudkøbing until he and his one year younger brother Anders in 1793 moved to Copenhagen. The teaching, the brothers got home, was distinctive. A German wig maker gave them their first lessons. The wig maker tried to inspire them with admiration for everything that was German, were looking for another teacher, a Norwegian student, to highlight Norway over Denmark in their eyes. But through their unusually large inquisitive, replaced by the widespread self-study in different directions, what was missing in traditional teaching. How well they followed with the passage of political and religious movements, portraying the younger brother living in Of my life and my time history. In 1794 they were both students, then after a brief preparation consisted second exam at the university with distinction in all subjects, and because of their kinship with the founder got the room on Elers College.

Ørsted havde fra han var 12 år været elev i farens apotek; kemien interesserede ham, og han læste alt hvad han kunne få fat i med kemisk indhold. Hans studie blev derfor farmakologi, og da han i 1797 tog farmaceutisk eksamen, imponerede han censorerne ved sin modenhed og dygtighed. Samtidig med at han læste til eksamen, besvarede han med held to af universitetets prisspørgsmål, det ene af æstetisk, det andet af kemisk indhold. Ligesom broren var Ørsted tidligt og stærkt påvirket af Immanuel Kant, hvilket bevidnes i hans doktordisputats (1799), en latinsk afhandling, oversat til dansk med titlen *Grundtrækkene af Naturmetafysikken*. Han følger her Kant næsten ord for ord, og opstiller det som den kritiske filosofis fordring, at "alle Naturlove bør udledes af vor Kendeevnes Natur". De her nævnte og flere mindre afhandlinger fra Ørsteds ungdom vidner om en livlig, alsidig interessere, men viser endnu ikke tegn på originalitet.

Early interest in chemistry

Ørsted had since he was 12 years been a student in his father's pharmacy; chemistry interested him, and he read everything he could get hold of chemical content. His study was therefore pharmacology, and when he took in 1797 his pharmaceutical exam, he impressed the examiners by his maturity and skill. While he was studying for the exam, he successfully answered two of the university's prize issues, one on aesthetics, the second of chemical content. Like his brother, Ørsted was early on heavily influenced by Immanuel Kant, as evidenced in his doctoral thesis (1799), a Latin treatise, translated into Danish entitled *The basics of Natural Metaphysics*. He follows here Kant almost word for word, and sets it as the critical philosophy's claim that "all laws of nature should be derived from our ability to know nature". Johannes Kepler acknowledged his Creator in a to the Bavarian chancellor in 1599, thusly:

Those laws [of nature] are within the grasp of the human mind; God wanted us to recognize them by creating us after his own image so that we could share in his own thoughts.

— Johannes Kepler

Letter (9/10 Apr 1599) to the Bavarian chancellor Herwart von Hohenburg. Collected in Carola Baumgardt and Jamie Callan, Johannes Kepler Life and Letters (1953), 50.

6. ALLESANDROS GALVANIC ELEMENT

Det, der bragte Ørsted ind på nye baner, var Voltas opdagelse af det galvaniske element i år 1800. I dette år havde Ørsted overtaget ledelsen af Løveapoteket under ejeren, professor Mantheys fravær. Derved fik han mulighed for selv at eksperimentere med Voltas søjle og studerede især dens kemiske effekter. For at måle disse konstruerede han i 1801 et apparat, der meget minder om Faradays voltameter (voltmeter). Han var ivrig efter at rejse for at møde udlandets store kemikere og fysikere, men samtidig søgte han et professorat i fysik der var blevet ledigt efter A.N. Aasheims død. Det fik han ikke, han måtte nøjes med d. 7. november 1800 at blive adjunkt ved universitetet uden løn. Under disse omstændigheder var der intet, der holdt ham hjemme, og i sommeren 1801 begyndte han sin første udenlandsrejse, finansieret af det cappelske legat på 500 rigsdaler årligt og 400 rigsdaler fra Fonden ad usus publicos.

The galvanic element

What brought Ørsted into new research, was Volta's discovery of the galvanic element in 1800. In that year, Ørsted had taken over the management of the Lion Pharmacy under absence of the owner, professor Manthey. Thus, he was able to experiment with Volta's piles and studied especially its chemical effects. To measure these, he constructed in 1801 an apparatus that was very reminiscent of Faraday volt meter. He was eager to travel to meet prominent foreign chemists and physicists, but at the same time he sought a professorship in physics, left vacant by A. N. Aasheim's death. That he did not, he had to settle for d. November 7, 1800 to become an assistant professor at the University without pay. Under these circumstances, there was nothing that kept him at home, and in the summer of 1801 he began his first trip abroad, funded by the cappelske grant of 500 dollars per year and 400 dollars from the fund for public use.

Oersted holds a wire above a magnetic needle supported on a pivot. The needle is deflected when electric current flows through the wire.

Oersted hält einen Draht über einer magnetischen Nadel, die auf einem Zapfen getragen wird. Die

Nadel wird abgelenkt, wenn elektrischer Strom durch den Draht fließt.

Oersted hålla en tråd över en magnetisk nål uppburen på en pivot. Nålen avlänkas när elektrisk ström flyter genom tråden.

Ørsted holde en tråd over en magnetisk nål understøttet på en drejetap. Nålen afbøjes, når elektrisk strøm gennem tråden.

Rejsen gik til Tyskland, hvor han i Weimar traf Johann Wilhelm Ritter, som han nærede et stort ønske om at møde. Ørsteds beundring for Ritter var velfortjent, men kun delt af få i hans samtid.

He traveled to Germany, where he met Johann Wilhelm Ritter in Weimar, for whom he had a great desire to meet. Ørsteds admiration for Ritter was well deserved, but only shared by few of his contemporaries.

Han reste till Tyskland, där var han i Weimar träffade Johann Wilhelm Ritter, som han hade en stor önskan att möta. Ørsteds beundran för Ritter var välförtjänt, men endast delas av några av hans samtida.

Il a voyagé en Allemagne, où il était à Weimar a rencontré Johann Wilhelm Ritter, qu'il avait un grand désir de se rencontrer. Ørsteds admiration pour Ritter était bien méritée, mais seulement partagé par quelques-uns de ses contemporains.

Ørsted knyttede straks venskab med ham, skaffede ham senere ved Franz Xaver von Baaders hjælp en stilling i München og arbejdede under sit ophold i Paris på at skaffe ham anerkendelse og udmærkelse, men uden held. Opholdet i Berlin var på mange måder interessant og lærerigt. Han kom meget i den kendte fru Henriette Herz' hus, hvor han foreviste nye elektriske forsøg og mødte mange store videnskabsmænd. Han hørte forelæsninger af Johann Gottlieb Fichte og Friedrich Schlegel og befandt sig i det hele taget her i omgivelser, der tiltalte ham.

Ørsted immediately made friends with him, and got him later with Franz Xaver von Baader's help a position in Munich and worked during his stay in Paris to give him recognition and honors, but without success. The stay in Berlin was in many ways interesting and instructive. He received much knowledge in the house of the well-known Mrs. Henriette Herz, where he demonstrated new electrical tests and met many eminent scientists. He heard lectures by Johann Gottlieb Fichte and Friedrich Schlegel and was, in general, in surroundings which appealed to him.

I München mødte han grev Rumford, men satte særlig stor pris på bekendtskabet med Franz Baader, om hvem han har sagt: "Han driver aldeles på, at den moralske og fysiske natur på det nøjeste hænge sammen, og at uden en sådan forbindelse fysikken egentlig ingen værd har. Han stemmer i denne Henseende meget med Ritter og jeg med begge". Det er lignende tanker som Ørsted senere forsøgte at videreføre.

7. MORALITY AND PHYSICAL NATURE ARE INTIMATELY RELATED

In Munich, he met Count Rumford, but greatly valued his acquaintance with Franz Baader, of whom he said: "He maintains that <u>moral and physical nature are closely interdependent, and that without such a connection physics really is not worth having.</u> He agrees in this respect both with Ritter and me." These are similar ideas which Ørsted tried to pursue later.

A Munich, il a rencontré le comte Rumford, mais grandement apprécié sa rencontre avec Franz Baader, dont il a dit: "Il affirme que la nature morale et physique sont étroitement interdépendants et sans cette connexion la physique est vraiment pas la peine d'avoir. Il est d'accord à cet égard avec Ritter et moi." Ce sont des idées similaires som Ørsted essayé deux poursuivre plus tard.

The oersted unit was established by the IEC in 1930 in honor of the Danish physicist Hans Christian Ørsted. Ørsted discovered the connection between magnetism and electric current when a magnetic field produced by a current-carrying copper bar deflected a magnetized needle during a lecture demonstration.

L'unité oersted a été créée par la CEI en 1930 en l'honneur du physicien danois Hans Christian Ørsted. Ørsted a découvert la connexion entre le magnétisme et le courant électrique lorsqu'un champ magnétique produit par une barre de cuivre porteur de courant a dévié une aiguille magnétisée lors d'une démonstration de conférence.

Oersted (abbreviated as Oe) is the unit of the auxiliary magnetic field H in the CGS system of units. It is identical to Dyne/Maxwell.

Oersted (abrégé Oe) est l'unité du champ magnétique auxiliaire H dans le système des unités CGS. Il est identique à Dyne / Maxwell.

Oersted (abgekürzt als Oe) ist die Einheit des Hilfsmagnetfeldes H im CGS-System von Einheiten. Es ist identisch mit Dyne / Maxwell.

Ørsted defineres som en dyn pr stang. Den Ørsted er $1000 / 4\pi$ (≈ 79.5774715) ampere per meter, i form af SI-enheder.

Ørsted defineres som en dyn pr stang. The Ørsted er $1000 / 4\pi$ (≈ 79.5774715) amp per meter, i form af SI-enheder.

H-feltstyrken i en lang solenoide såret med 79,58 omdrejninger pr meter af en ledning transporterer 1 A er cirka 1 ørsted. Den foregående redegørelse er præcis korrekt, hvis solenoiden overvejes, er uendelig længde med den nuværende jævnt fordelt over dens overflade.

The H-field strength for a long solenoid wound with 79.58 turns per meter of a conductor carrying 1A is approximately 1 Oe. The foregoing disclosure is exactly correct if the solenoid is considered of infinite length with the current spread evenly over its surface.

The magnetic field B, n is the number of turns per unit length, sometimes called the "turns density". The expression is an idealization to an infinite length solenoid, but provides a good approximation to the field of a long solenoid.

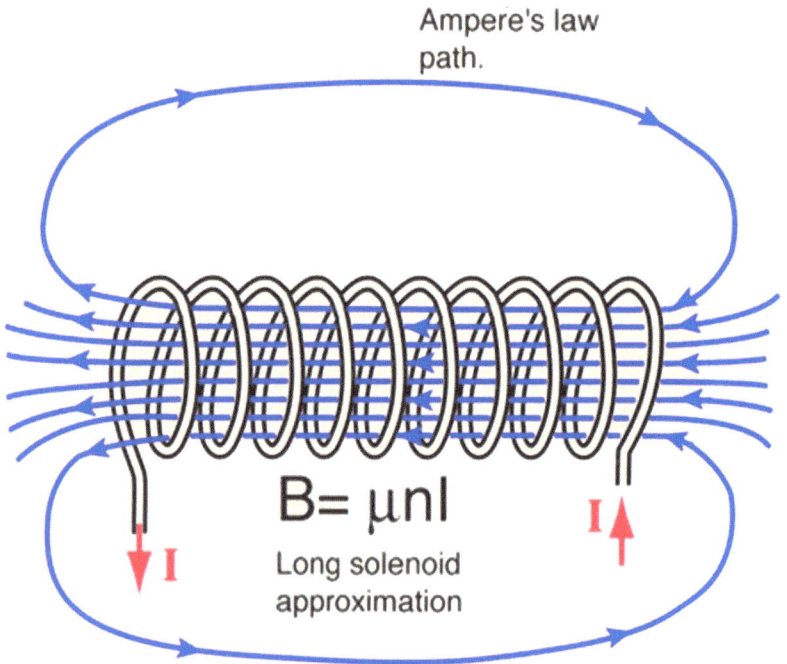

Ampere's law path.

$B = \mu n I$

Long solenoid approximation

The magnetic field is concentrated into a nearly uniform field in the center of a long solenoid. The field outside is weak and divergent.

Magnetic field = permeability x turn density x current

For a solenoid of length L = $\boxed{0.5}$ m with N = $\boxed{100}$ turns, the turn density is n=N/L= $\boxed{200}$ turns/m.

If the current in the solenoid is I = $\boxed{1}$ $\boxed{1}$ amperes

and the relative permeability of the core is k = $\boxed{200}$,

then the magnetic field at the center of the solenoid is

B = $\boxed{0.5026}$ Tesla = $\boxed{502.655}$ gauss.

The Earth's magnetic field is about half a gauss.

The relative permeability of magnetic iron is around 200

http://hyperphysics.phy-astr.gsu.edu/hbase/magnetic/solenoid.html

Ørsted er nært beslægtet med den gauss, CGS enhed af magnetiske fluxtæthed. I et vakuum, hvis magnetisering feltstyrke er en Oe, så magnetfeltet tæthed er en G, der henviser til, et medium, der har permeabilitet r (i forhold til permeabilitet vakuum).

Oe is closely related to the gauss, the CGS unit of magnetic flux density. In a vacuum, where the magnetization field strength is one Oe, so that the magnetic field density is one G, that indicates a medium having permeability s (compared to the permeability of vacuum).

I december 1801 kom han til Paris, hvor han blev et helt år. Han hørte forelæsninger af Louis Nicolas Vauquelin, Antoine François de Fourcroy, Louis Jacques Thénard og Claude Louis Berthollet over kemi, af Jacques Charles og Jean-Baptiste Biot over fysik, af Georges Cuvier over naturhistorie. Han så, at han kunne lære meget her, selvom han var forudindtaget imod det franske videnskabelige samfund, som med sin mere eksakte matematiske karakter stod i skarp kontrast til den naturfilosofiske tankegang, og selvom han tit mærkede, hvor lidt man i Paris anerkendte, hvad der blev udrettet i andre lande; han måtte finde sig i at blive spurgt om han kendte en platindigel og om han havde set et galvanisk element. Men tiden benyttede han på bedste måde ved at høre forelæsninger, se fabrikker og søge personligt bekendtskab med de største videnskabsfolk.

In December 1801, he moved to Paris, where he stayed a full year. He heard the lectures of Louis Nicolas Vauquelin, Antoine François de Fourcroy, Louis Jacques Thénard and Claude Louis Berthollet, the chemistry of Jacques Charles and Jean-Baptiste Biot, and the physics of Georges Cuvier of natural history. He saw that he could learn a lot here, even though he was biased against the French scientific community, which with its more exact mathematical character. This was in sharp contrast to the natural philosophical thinking, and although he often noticed how little was acknowledged in Paris of the work done other countries. He would find himself being asked if he knew what a platinum crucible was and if he had seen a galvanic element. But he used the time there in the best way by hearing lectures, observing factories and seeking personal acquaintance with the most renowned scientists.

Ved sin hjemkomst i januar 1804, håbede Ørsted at blive professor i fysik, men man så skævt til ham på grund af hans interesse for naturfilosofien;

dog blev han tildelt en lønning på 300 rigsdaler i tre år og lige så meget til eksperimenter. Han fik overladt en samling apparater, som havde tilhørt dr. Thomas Buntzen, men måtte selv leje lokale til forelæsningerne. Disse drejede sig mest om kemisk fysik og var så stærkt besøgte, også af kvinder, at han knap kunne skaffe plads til tilhørerne. Foruden at virke som lærer, udfoldede han i disse år en ikke ubetydelig forfattervirksomhed, mere eller mindre i naturfilosofisk retning.

Specialundersøgelsen interesserede ham mindre, generelle synspunkter, endog vovelige spekulationer tiltalte ham mere. For ham var spørgsmålet ikke: "Hvilke er de enkelte syrers eller basers egenskaber?" men "Hvorfor neutraliserer syrer og baser hinanden?" eller "Hvorfor må der tilsættes noget til vandet for at opløse et metal?" Overalt, hvor der blot fandtes en antydning af en besvarelse af disse sidste spørgsmål, anede han noget betydningsfuldt. Af hans afhandlinger fra den periode skal vi fremhæve *Betragtninger over Kemiens Historie*, indledningsforedrag holdt i vinteren 1805-6 (Samlede og efterladte Skrifter V, 3), i hvilke Ørsteds særlige stil og tankegang for første gang træder frem.

On his return in January 1804 Ørsted hoped to become a professor of physics, but was not taken seriously because of his interest in natural philosophy. However, he was awarded a salary of 300 dollars for three years and as much for experiments. He had been left a collection of apparatus that had belonged to Dr. Thomas Buntzen, but he even had to rent rooms for his lectures. These mostly concerned chemical physics and were heavily attended, also by women, so he could barely make room for the audience. In addition to serving as a teacher, he developed as a writer in later years, in the natural philosophical direction. Specialized study interested him less, general views, even daring speculations appealed to him more. For him, the question was not: "What are the characteristics of various acids or bases?", but "Why do acids and bases neutralize each other?", or "Why cannot add anything be added to water to dissolve a metal?" Wherever there just was a hint of an answer to the last question, he did something significant. Of his theses from the period, we highlight *Reflections on History of Chemistry*, introductory lectures held in the winter of 1805-6 in which Ørsted's style and way of thinking emerges for the first time.

Under et ophold i Berlin udarbejdede Ørsted et af sine vigtigste skrifter, *Ansichten der chemischen Naturgesetze* (1812). Først gives en oversigt over de vigtigste stoffer og processer, dernæst behandles de elektriske kræfter fra et kemisk synspunkt, endelig anføres de grunde, der taler for, at alle legemer indeholder elektriske kræfter, som dog ikke træder åbenlyst frem, eftersom de holder hinanden i ligevægt. Han antager, at de elektriske kræfter udbreder sig ved svingninger, som fremkaldes ved fordeling. Eftersom svingningerne møder større eller mindre modstand, vil der opstå en tilsvarende varme. Lyset opstår ved elektriske svingninger i dårlige ledere. Også forholdet mellem elektricitet og magnetisme behandles her. Ørsted stod stort set alene med disse anskuelser, og de vandt heller ingen anerkendelse, da de jo ikke hvilte på noget almindelig anerkendt grundlag, og Ørsted selv evnede ikke at udvikle dem således, at deres betydning derigennem kunne blive forstået. Det blev først langt senere muligt at vinde betydningsfulde resultater ad denne vej; og den senere eksakte naturforskning arbejder netop ud fra disse synspunkter, og udtrykker sig med de samme ord. Men Ørsted kunne dog, da han otte år senere opdagede elektromagnetismen, henføre til, at den afgav en bekræftelse på hans naturopfattelse.

During a stay in Berlin drew Ørsted one of his most important writings, *Ansichten der Chemischen Naturgesetze* (1812). Firstly, an overview of the main substances and processes, and he considered electric power from a chemical point of view, finally, he suggested the reasons that all bodies contain electric forces openly appear, since they are in balance with each other. He assumed that electrical forces propagate by vibrations induced by the distribution. Since the oscillation meet more or less resistance, there will be a corresponding heat. The light generated by electrical oscillations in poor conductors. Also, he investigated the relationship between electricity and magnetism. Ørsted stood virtually alone with these views, and they achieved no recognition, since they were not based on any scientific consensus. Ørsted himself did not to develop them so that their importance was unable to be understood. It was not until much later possible to gain significant conclusions from his research. But Ørsted could, however, when he discovered electromagnetism eight years later, confirm his view of nature.

Angående denne opdagelse har Ørsted selv udtalt følgende: Under en række forelæsninger, han holdt i foråret 1820, blev tanken om sammenhængen mellem elektricitet og magnetisme særlig levende hos ham; han sendte da en stærk udladning gennem en fin platintråd og så virkelig, at en magnetnål, som befandt sig i nærheden, derved kom i bevægelse.

For this discovery, Ørsted himself stated the following: In a series of lectures he delivered in spring 1820, the idea of the relationship between electricity and magnetism was especially crucial with him; he sent a strong discharge through a fine platinum wire that was so effective that a magnetic needle in the area was visibly deflected.

Men det var vanskeligt for ham at forstå virkningens natur, og han udsatte derfor den nærmere undersøgelse, indtil han havde skaffet sig et kraftigt galvanisk batteri. Han genoptog derefter forsøgene i juli måned. 21. juli 1820 udsendte han en kort latinsk beretning om forsøgene. Kun fire kvartsider stor indeholder den dog en stor mængde forsøg, som viser, at her virkelig er tale om en vekselvirkning mellem de to naturkræfter, mens legemerne, der bærer dem, og omgivelserne er uden indflydelse.

It was difficult for him to understand the impact; nature, and he therefore suspended further investigation until he had obtained a powerful galvanic battery. He resumed then the trials in July. On July 21, 1820, he released a short report in Latin on the trials. It detailed many experiments that showed that this was indeed an interaction between the two forces of nature, while the bodies that carry them, and the surroundings did not influence the results.

Betydningen af Ørsteds opdagelse stod straks klar for alle sagkyndige, og den blev udgangspunkt for den række opdagelser, som knytte sig til André-Marie Ampères og François Aragos navne, og som 11 år senere afsluttedes ved Faradays opdagelse af induktionen.

The importance of Oersted's discovery was immediately clear to all experts, and it became the basis for the series of discoveries that are associated with the names of André-Marie Ampère and François Aragos, and ended 11 years later with Michael Faraday's discovery of induction.

Er riecht die Wahrheit, He [Faraday] smells the truth.
— Friedrich Wilhelm Georg Kohlrausch, Quoted in John Tyndall, *Faraday as a Discoverer* (1868), 45.

His contemporaries felt how much they were indebted to Ørsted, there is testimony enough about that, and many academic honors were awarded to him from many admirers. Although it largely was others who drew upon the consequences of his discovery, it should be emphasized that Ørsted showed that the law of action and reaction applies to electromagnetism. On the other hand, he failed to demonstrate the influence of earth's magnetism on an electric current, stating that his lab apparatus lacked sufficient sensitivity.

Ørsteds regel – hold højre hånd med fingerspidserne i strømmens retning (plus→minus). Højre håndflade skal være mellem kompasset og ledningen. Kompassets magnetnordpol vil da slå ud til tommelfingersiden.

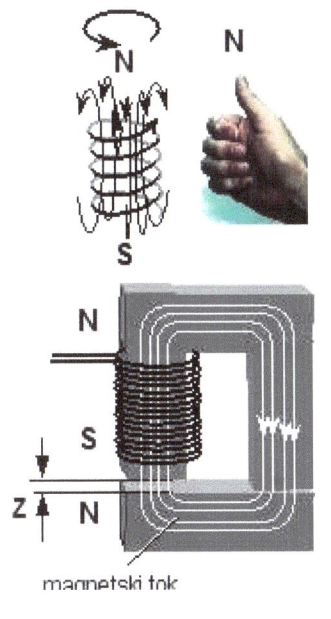

Ørsteds rule - holding the right hand with fingers in the direction of flow (plus → minus). Right palm should be between the compass and the cord. Compass magnetic north pole will then turn out to the thumb side.

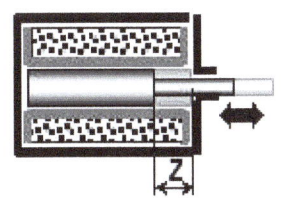

magnetski tok

Fra de sidste tyve år af Ørsteds liv findes ingen større videnskabelige arbejder fra hans hånd, men mange meddelelser, ofte af original karakter, vidner om den interesse og forståelse, hvormed han fulgte videnskabens fremgang. Derom vidner også den levende korrespondance, han førte med udlandets lærde. Det personlige forhold til disse holdt han ved lige ved gentagne udenlandsrejser, også i hans høje alder. Herunder må desuden nævnes hans ivrige deltagelse i de nordiske naturforskermøder, hvor han naturligt indtog en meget fremtrædende plads.

Efter at vi nu har gennemgået hovedpunkterne i Ørsteds videnskabelige arbejde, skal vi omtale andre beslægtede sider af hans virksomhed.

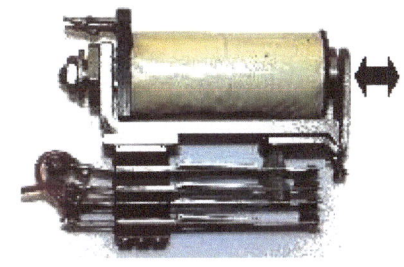

From the last twenty years of Ørsted's life were no greater scientific works from his hand, but a lot of messages, often of original character, testifying to the interest and understanding with which he followed the progress of science. He maintained personal relationships by repeated trips abroad, even in his old age. Below should also mention his eager participation in the Nordic naturalist meetings where he naturally occupied a very prominent place.

Ørsted, som endelig d. 18. juli 1817 blev udnævnt til professor ordinarius ved universitetet, var til stadighed interesseret i at udbrede kendskab til naturvidenskaben, også uden for de studerendes kreds. I den forbindelse udsendte han 16. oktober 1823 en opfordring til at stifte et selskab med dette formål. Selvom sagen næppe fandt den tilslutning, som Ørsted havde ventet, konstitueredes Selskabet for Naturlærens Udbredelse d. 26. marts 1824 med daværende kronprins Christian Frederik, senere kong Christian 8., som "patron". Dette selskab, som stadig eksisterer, skulle ifølge Ørsteds oprindelige idé desuden støtte industrielle virksomheder og forskning samt støtte unge, der søgte teknisk uddannelse. Ørsted blev, som naturligt var, selskabets "bestandige" direktør. Der manglede dog både menneskelige og financielle resourcer til at gennemføre Ørsteds plan. Dette blev dog delvis afhjulpet ved oprettelsen af den polytekniske læreanstalt. Georg Frederik Krüger Ursin indgav 1827 til kongen en ansøgning om oprettelse af en polyteknisk skole. Sagen blev forelagt universitetets til udtalelse; det i den forbindelse nedsatte udvalg, i hvilket Ørsted var sjælen, foreslog at give den påtænkte skole en mere videnskabelig karakter end oprindelig påtænkt. Efter at de store økonomiske vanskeligheder var overvundet, stiftedes læreanstalten d. 27. januar 1829. Ørsted blev dens første direktør, og ham tilskrives en stor del af den anseelse, den efterhånden vandt, og den ånd, hvori den er ledet.

Ørsted, was appointed professor ordinarius at the university on 18 July 1817, was constantly interested in spreading knowledge of science, also beyond the circle of his students. In this regard, he issued October 16, 1823 an invitation to form a company for this purpose. Although the case hardly found the connection that Ørsted had expected, was constituted the Society for the Dissemination of Natural Science on 16 March 1824, with the then Crown Prince Christian Frederik, later King Christian 8, as "patron". This company, which still exists, was envisioned to advance Ørsted's original ideas and support industrial companies and research, as well as support young people who were looking for technical training. Ørsted was the company "permanent" director. It lacked, however, both human and financial resources to implement Ørsteds plan. This was partially remedied by the creation of the Polytechnic. Georg Frederik Krüger Ursin submitted in 1827 to the King an application for the establishment of a polytechnic school. The case was submitted for the university's decision; in this context, a committee established in which Ørsted was the soul. It proposed giving the proposed school a more scientific character than originally planned. After great financial difficulties were overcome, it was founded on 27January 1829. Ørsted was its first director, and a large part of the reputation it eventually won, and the spirit in which it is headed, was attributed to him.

Ørsteds lectures were heavily influenced by his whole personality and his highlight of spirit and nature's unity. They had a major attraction on the audience. His original plan to provide a comprehensive presentation of science was interrupted when he entered his experimental work. After he had discovered electromagnetism, he developed this aspect of science so thoroughly and rapidly that it was difficult to describe adequately. Only the mechanical aspect of the natural sciences, which he completed and published many times in different formats, the first time in 1809, and later by Carl Valentin Holten in 1859, was contained in a detailed introductory section.

Natural science, does not simply describe and explain nature; it is part of the interplay between nature and ourselves.
— Werner Heisenberg, In Physics and Philosophy: The Revolution in Modern Science (1962), 81.

8. ØRSTEDS VERDENSOPFATTELSE-OERSTED'S WELTANSCHAUUNG

Hans Christian Ørsted, *Der Geist in der Natur*, 1854

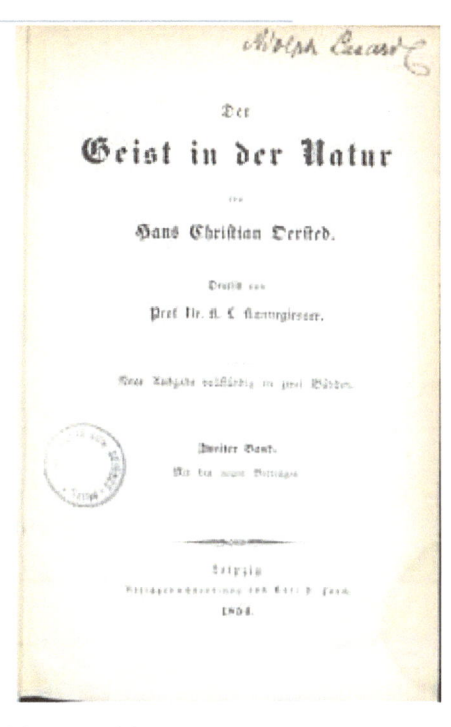

Ligesom sin bror havde H. C. Ørsted en varm interesse for alle livets sider. Hans studie af naturen var derfor ikke kun indskrænket til fordybelsen i konkrete enkeltsager; men ligesom han med forkærlighed dvælede ved de store almene grundspørgsmål, udvidede hans granskning sig efterhånden uvilkårlig til at skabe en hel naturfilosofi, eller endda et helt tilværelsessystem, hvori intet væsentligt var oversprunget.

Like his brother, H. C. Ørsted had a passionate interest in all aspects of life. His study of nature was not only restricted to the immersion in specific individual cases; but he dwelt with fondness on great general fundamental questions, extending his analysis gradually to create an entire philosophy of nature, or even an entire system of existence, overlooking nothing.

Hele hans tidsånd havde retning mod det altomfattende; den romantiske naturfilosofi havde vundet indpas i Tyskland, og i hans tidlige arbejde var Ørsted også til dels blevet tiltrukket af denne retning. Men det gik med ham som med Adam Oehlenschläger og F.C. Sibbern: der var for meget sandhed og sundhed i det danske samfund allerede dengang, til at retningen med alle dens udskejelser kunne vinde indpas her. Ørsted blev aldrig naturfilosof i tysk forstand, men lærte hurtigt at skelne mellem, hvor langt tanken virkelig kunne føre os, og hvor den subjektive fantasi begyndte. Han fremhævede konstant med stor bestemthed, at vi altid måtte begynde med iagttagelse og forsøg og om muligt også helst efterprøve de følgende resultater på samme måde, før det kunne afgøres som faktum.

His whole *Zeitgeist* tended toward the all-encompassing; this romantic natural philosophy had gained ground in Germany, and in his early work Ørsted was also partly been attracted to this direction. But it went with him as with Adam Oehlenschläger and F.C. Sibbern: there was too much truth and health in Danish society even then, for it to could gain a foothold here. Ørsted was never natural philosopher in the German sense, but quickly learned to distinguish how far the idea could really take us and where the subjective imagination began. He emphasized constantly with great certainty that we always had to begin with observation and experimentation and, where possible any time verify the following results, before it could be established as fact.

Det var den excentriske tyske forsker Ritter, der delagtiggjorde Ørsted i den naturanskuelse, at naturkræfterne ikke er adskilte og særprægede enheder knyttede til enkeltstående fænomener, men at de tværtimod tilsammen udgør en afrundet helhed, forde de alle i sidste ende hidrører fra samme skabende kraft. Naturens enhed er ikke som et mekanisk system, men som en levende organisme, der på mangfoldig vis er forbundet med den menneskelige virksomhed. Mennesket selv er et væsen fra naturen og underkastet dens love, blot med den forskel, at mennesket har bevidsthed herom.

It was the eccentric German scientist Ritter, who shared with Ørsted in the conception of nature that natural forces are separate and distinctive units attached to individual phenomena, but on the contrary, make up a rounded whole, because they all are ultimately derived from the same creative force. Nature's unity is not a mechanical system, but a living organism, which in many and various ways are associated with human activity. Man, himself, is a creature of nature and subject to its laws, with the only difference that human beings have consciousness.

9. NIKOLA TESLA - THE GREATEST SCIENTIFIC GENIUS OF THE 20TH CENTURY

Wissenschaftliche Studien, die beweisen, dass Bewusstsein und die physische Welt einander beeinflussen - Scientific studies that prove that consciousness and the physical world influence each other!

Tesla formulierte es am besten: „Ab dem Tag, an dem die Wissenschaft beginnt, nicht-physikalische Phänomene zu analysieren, werden in einem Jahrzehnt mehr Fortschritte gemacht werden als in all den Jahrhunderten seit deren Bestehen. Um die wahre Natur des Universums zu verstehen, muss man in Begriffen wie Energie, Frequenz und Vibration denken. "

Tesla put it best: "From the day when science begins to analyze non-physical phenomena, more progress will be made in a decade than in all the centuries since its existence. To understand the true nature of the universe, one must think in terms of energy, frequency, and vibration. "

Swami Vivekananda, ein indischer Hindu-Mönch und wichtigster Schüler des heiligen Ramakrishna im 19. Jahrhundert, war Teslas Mentor. Er sagt: Wissenschaft funktioniert am besten, wenn sie in Harmonie mit der Natur ist. Wenn wir diese beiden zusammenbringen, können wir Entdeckungen zu grossartigen Technologien, wie zum Beispiel Freie Energie, machen, die sich nur ergeben, wenn das Bewusstsein des Planeten bereit ist, sie anzunehmen.

Swami Vivekananda, an Indian Hindu monk and most important disciple of Ramakrishna in the 19th century, was Tesla's mentor. He says: Science works best when it is in harmony with nature. When we bring these together, we can make discoveries to great technologies, such as free energy, which only arise when the planet's consciousness is ready to accept them.
https://nebadonia.wordpress.com/2016/02/04/wissenschaftliche-studien-die-beweisen-dass-bewusstsein-und-die-physische-welt-einander-beeinflussen/

Wie Nikola Tesla „Spiritualität" und Philosophie nutzte, um über die Realität, unbegrenzte Energieressourcen und Wissenschaft zu lernen.
How Nikola Tesla used "spirituality" and philosophy to learn about reality, unlimited energy resources and science.

Die Eigenschaften des Raumes – The Properties of Space

Die Wissenschaft funktioniert am besten, wenn sie in Einklang mit der Natur ist. Wenn wir diese beiden zusammenfügen, können wir grossartige Technologien entdecken, die nur dann zustande kommen können, wenn das Bewusstsein des Planeten bereit ist, sie zu integrieren. Ein Beispiel dafür ist die „Freie Energie", auch bekannt als „Nullpunktenergie", welche die feinstoffliche Substanz ausnutzt, die um uns herum existiert und sie in nutzbare Energie wandelt. Dies würde uns eine unbegrenzte Energiequelle zur Verfügung stellen und würde praktisch alle Armut auf dem Planeten eliminieren.

Science works best when it is in harmony with nature. When we combine these two, we can discover great technologies that can only happen when the planet's consciousness is ready to integrate them. An example of this is "free energy", also known as "zero-point energy", which exploits the fine substance that exists around us and transforms it into usable energy. This would provide us with an unlimited energy source and would virtually eliminate all poverty on the planet.

Die Eigenschaften des Raumes wurden von vielen postuliert, wir finden sie in der alten vedischen Philosophie, den östlichen Mysterien, verschiedenen alten Zivilisationen der Geschichte der Menschheit … bis hin zu Descartes, Einstein, Newton und vielen anderen. Die Menschen sind neugierige Wesen, und unsere Suche nach dem „Entdecken, was ist" wird nie zu Ende gehen.....

The properties of space were postulated by many, we find them in ancient Vedic philosophy, the Eastern Mysteries, various ancient civilizations of the history of mankind ... to Descartes, Einstein, Newton and many others. People are curious beings, and our quest for "discover what is" will never end ...

Lost secrets

After Nikola Tesla died, there was a scramble by the United States government to find all his papers, notes and research before other foreign powers could find them. Tesla's nephew, Sava Kosanovic, reported someone had obviously gone through Tesla's belongings and took *an unknown number of personal notes and papers*.

What happened with **forgotten papers** of Nikola Tesla?

It was known by the FBI that German intelligence had already spirited away a *sizable amount of Tesla's research* several years before his death.

This stolen material, it is thought, would eventually result in the development of the *Nazi flying saucer*.

Es wurde durch das FBI bekannt, dass deutsche Nachrichtendienste hat, schon mehrere Jahre vor Teslas Tod, eine beträchtliche Menge an Teslas Forschung weggezaubert. Es wurde angenommen, dass diese gestohlene Materialen schließlich in der Entwicklung der Nazi-fliegende Untertasse resultieren würde.

Det var kendt af FBI at tyske efterretningstjeneste havde allerede stjålet et betydeligt beløb af Teslas forskning, flere år før hans død. Dette stjålne materiale, det menes, ville i sidste ende resultere i udvikling af den nazistiske flyvende tallerken.

The United States made sure that this would not happen again. Anything even remotely associated with the great acientist was quickly confiscated and lost within the secret networks of pre-World War II America.

Nevertheless, more than a dozen boxes of Tesla's belongings left behind at hotels like the **Waldorf Astoria**, the **Governor Clinton Hotel** and the **St. Regis** had already been sold to salvagers to pay off Tesla's outstanding bills. **Most of these boxes and the secrets they contained have never been found.**

Later in life Tesla made claims concerning a "teleforce" weapon, after studying the Van de Graaff generator. The press referred to it as a "peace ray" or death ray. Tesla described the weapon as capable of being used against ground-based infantry or for anti-aircraft purposes. Tesla gives the following description concerning the "particle gun's operation:

Senare i livet Tesla gjorde påståenden om en "teleforce" vapen, efter att han har studerat Van de Graaffs generator. Pressen hänvisade till det som en "fredens ray" eller dödens strål. Tesla beskrev vapen som kan användas mot markbaserad infanteri eller för luftvärnsändamål. Tesla gav följande beskrivning om "partikelkanons verksamhet:

"The nozzle would send concentrated beams of particles through the free air, of such tremendous energy that they will bring down a fleet of 10,000 enemy airplanes at a distance of 200 miles from a defending nation's border and will cause armies to drop dead in their tracks."

"Munstycket skulle skicka koncentrerade strålar av partiklar genom den fria luften, sådan enorm energi att de kommer att få ner en flotta av 10.000 fientliga flygplan på ett avstånd av 200 miles från en försvarande nations gräns och kommer att orsaka arméer att släppa döda i sina spår . "

Tesla's teleforce weapon's components and methods included:

1. An apparatus for producing energy in free air instead of in a high vacuum as in the past.
2. A mechanism for generating a tremendous electrical force.
3. A means of intensifying and amplifying the force developed by the second mechanism.
4. A gun for producing a tremendous electrical repelling force. Tesla claimed to have worked on plans for a directed-energy weapon from the early 1900s he died.

In 1937, at a luncheon in his honor concerning the death ray, Tesla stated, "But it is not an experiment ... I have built, demonstrated and used it. Only a little time will pass before I can give it to the world." His records indicate that the device is based on a narrow stream of small tungsten pellets that are accelerated via high voltage, similar to his magnifying transformer.

Tesla wrote a treatise, *The New Art of Projecting Concentrated Non-Dispersive Energy through the Natural Media*, concerning charged particle beam weapons. Tesla naively hoped that his "superweapon ...would put an end to all war." This treatise is currently in the Nikola Tesla Museum archive in Belgrade. It describes an open-ended vacuum tube with a gas jet seal that allows particles to exit, a method of charging particles to millions of volts, and a method of creating and directing non-dispersive particle streams (through electrostatic repulsion). Tesla tried to interest the US War Department, the United Kingdom, the Soviet Union, and Yugoslavia in the device.

Tesla said that attempts were made to steal this invention, but the spies left empty-handed, since the design for the teleforce weapon was only contained in his head.

Tesla's lost journals revealed that while he was in Colorado Springs in 1899, he had intercepted communications from **extraterrestrial beings** who were secretly controlling mankind. These creatures were slowly preparing humans for eventual conquest and domination, using a program that had been in place since the creation of humankind, but was now accelerating due to the increase of scientific awareness on the earth.

Tesla described his years of research to interpret the strange radio signals, and how he attempted to inform the government and military concerning what he had learned, but his letters apparently went unanswered.

Tesla skrev om sine mange års forskning at fortolke den mærkelige radiosignal, og hans forsøg på at underrette regeringen og militæret vedrørende hvad han havde lært, men hans breve gik tilsyneladende ubesvarede.

Tesla spoke in confidence to several of his benefactors, including *Colonel John Jacob Astor*, who owned the *Waldorf Astoria hotel*. These benefactors listened to Tesla and secretly funded what was to be the start of mankind's first battle to regain control of its own destiny - a battle set in motion by Tesla. Unfortunately, Tesla's financier John Jacob Astor was drowned with the Titanic!"

John Jacob Astor
Owner of the Waldorf Astoria hotel

Tesla talte i tillid til at Flere af hans velgørerer, herunder oberst John Jacob Astor, som ejede Waldorf Astoria Hotel. Disse benefaktorer lyttede til Tesla og hemmeligt finansieret hvad hvad man skal være starten på menneskehedens første kamp for at genvinde kontrollen over sin egen skæbne - en kamp sat i gang af Tesla. Desværre blev Teslas finansmand John Jacob Astor druknede med Titanic! "

http://electrical-engineering-portal.com/what-happened-with-forgotten-papers-of-nikola-teslala.

When Tesla died, the government took all his papers and classified them. They did return most of those papers to Tesla's country of birth, but the U.S. Government withheld some of those papers to hide certain technology from the public.

Da Tesla døde, regeringen tog alle hans papirer og klassificeret dem. De gav tilbage de fleste af disse papirer til Teslas fødeland, men det ser ud at den amerikanske regering tilbageholdt nogle af disse papirer til at skjule visse teknologi fra offentligheden.

http://peswiki.com/directory:wireless-transmission-of-electricity

http://electrical-engineering-portal.com/what-happened-with-forgotten-papers-of-nikola-tesla

10. TESLA'S SCALAR WEAPONS OF MASS DESTRUCTION

In 1952, Tesla's remaining papers and possessions were released to Sava Kosanovic and returned to Belgrade, Yugoslavia where a museum was created in the inventor's honor. For many years, under Tito's communist regime, it was extremely difficult for Western journalists and scholars to gain access to the Tesla archive in Yugoslavia; even then they could see only selected papers. This was not the case for Soviet scientists who came in delegations during the 1950s. Concerns increased in 1960 when Soviet Premier Khrushchev announced to the Supreme Soviet that "a new and fantastic weapon was in the hatching stage."

I 1952 blev Tesla resterende papirer og ejendele frigivet til Sava Kosanovic og vendte tilbage til Beograd, Jugoslavien, hvor et museum, der skabes i opfinderens ære. I mange år, under Titos kommunistiske styre, var det yderst vanskeligt for vestlige journalister og forskere at få adgang til Teslas arkiv i Jugoslavien; selv da de kunne se kun udvalgte dokumenter. Dette var ikke tilfældet for sovjetiske videnskabsmænd, der kom i delegationerne i 1950'erne. Bekymringer øget i 1960 da Sovjet Premier Khrusjtjov meddelt den Øverste Sovjet, at "en ny og fantastisk våben, som var i klækningsfase."

Mankind must put an end to war, or war will put an end to mankind. War will exist until that distant day when the conscientious objector enjoys the same reputation and prestige that the warrior does today. Never have the nations of the world had so much to lose, or so much to gain. Together we shall save our planet, or together we shall perish in its flames.

Menneskeheden må sætte krig at ende til, eller krig vil være til ende på menneskeheden. Krig vil eksistere indtil gjorde militærnægter nyder samme anseelse og prestige gjorde kriger gør i dag. Aldrig har verdens nationer havde så meget at tabe, eller så meget at vinde. Sammen skal vi redde vores planet, eller sammen skal vi omkomme i dens flammer.

Mänskligheten måste sätta krig att avsluta, eller krig kommer att sätta i slutet för mänskligheten. Krig kommer att existera tills gjorde avlägsen dag när vapenvägrare har samma rykte och prestige gjorde krigare gör idag. Aldrig har världens nationer hade så mycket att förlora, eller så mycket att vinna. Tillsammans ska vi rädda vår planet, eller tillsammans skall vi förgås i dess lågor.

- John Fitzgerald Kennedy
http://www.abovetopsecret.com/forum/thread16978/pg1

Despite President Kennedy's warning, the superpowers continued to develop countless weapons of mass destruction during the Cold War, and employed Tesla's discoveries, which were planned to improve the lives of people, not to destroy mankind.

Trods præsident Kennedys advarsel, de supermagter Sovjetunion og Amerika fortsatte med at udvikle utallige masseødelæggelsesvåben under den kolde krig, og ansat Teslas opdagelser, som blev planlagt til at forbedre livet for mennesker, ikke for at ødelægge menneskeheden.

The Soviet Union used Tesla's designs and documents to create super weapons. As soon as his new super weapons were deployed and ready, Khrushchev did "something dramatic", to compensate for his loss of face in 1961, when he had to withdraw Soviet missiles from Cuba. The U.S.S. Thresher, SSN 593, was the most advanced attack submarine of its time.

It was faster and quieter than any submarine ever built, and able to dive deeper than any submarine in the world. SSN 593 was considered the most advanced weapons system of its day, created specifically to seek out and destroy Soviet submarines.

SSN 593, qui considéré comme le système sur les armes les plus avancées de l'époque, créé spécifiquement pour rechercher et détruire les sous-marins soviétiques.

Den atomubåd SSN 593, som betragtes af de mest avancerede våben system af sin dag, var skabt specielt til at opsøge og ødelægge sovjetiske ubåde.

Den atomubåt SSN 593 som anses vara av de mest avancerade vapensystem av sin tid, som skapats speciellt för att söka upp och förstöra sovjetiska ubåtar.

Das Atom-U-SSN 593 wurde die modernsten Waffensysteme seiner Zeit betrachtet, besonders sowjetische U-Boote geschaffen zu suchen und zu zerstören.

Атомная подводная лодка SSN 593 считается наиболее современные системы вооружения своего дня, специально созданное для поиска и уничтожения советских подводных лодок.

Its new sonar (both passive and active) could detect other submarines and ships at greater range, and it was intended to launch the U.S. Navy's newest anti-submarine missile, SUBROC.

On April 10, 1963, the Soviets destroyed the U.S.S. Thresher with one of their new weapons. Scalar EM (electrogravitational) beams, focused through the ocean to interfere on the Thresher under the surface, recreated spurious EM energy in the sub's electrical control circuits, jamming them so that the sub lost control, sank to crush depth, and imploded. The U.S. nuclear-powered submarine Thresher sank 220 miles east of Boston. All 129 men aboard were lost.

These were the scalar beam weapons invented in 1904 by Nicola Tesla. Since he died in 1943, many nations have secretly developed his beam weapons which are now further developed and more devastating. Using a satellite one can: cause a nuclear like destruction; earthquake; hurricane; tidal wave; cause instant freezing - killing every living thing instantly over many miles; cause intense heat like a burning fireball over a wide area; and induce hypnotic mind control over a whole population; or remotely read anybody's mind on the planet.

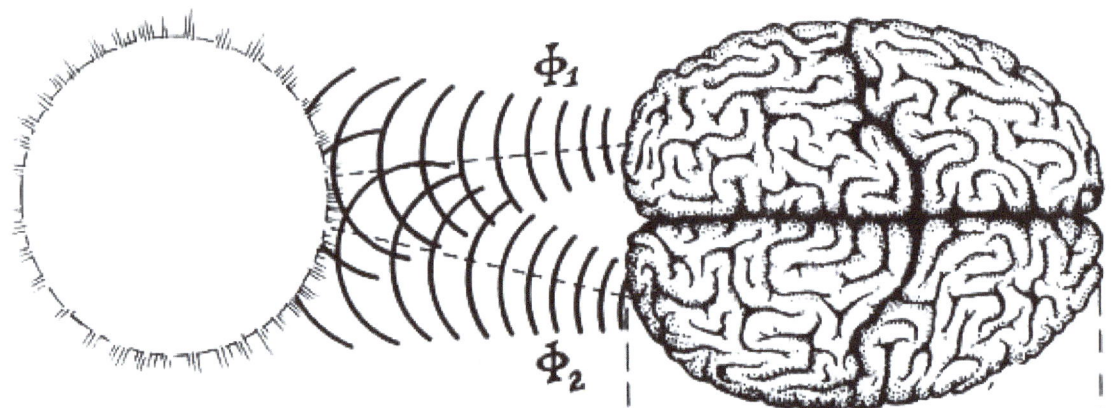

The human cerebral cortex is a natural scalar interferometer

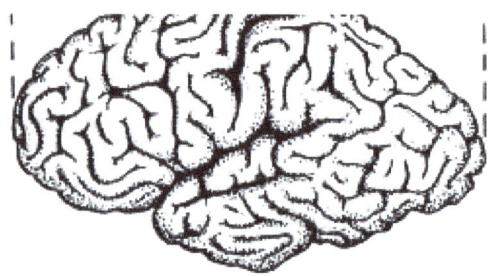

[Note: The human cerebral cortex is a natural scalar interferometer. It is a virtual state tuner, processor, and transmitter-receiver. It also can produce ... and control to some extent ... phase conjugate energy and phase conjugate waves. http://www.cheniere.org/books/aids/kindlingetc.html]

It can also be used to affect anybody's REM dream sleep by sending in subliminal pictures to the visual cortex; **cause hallucinogen drug like**

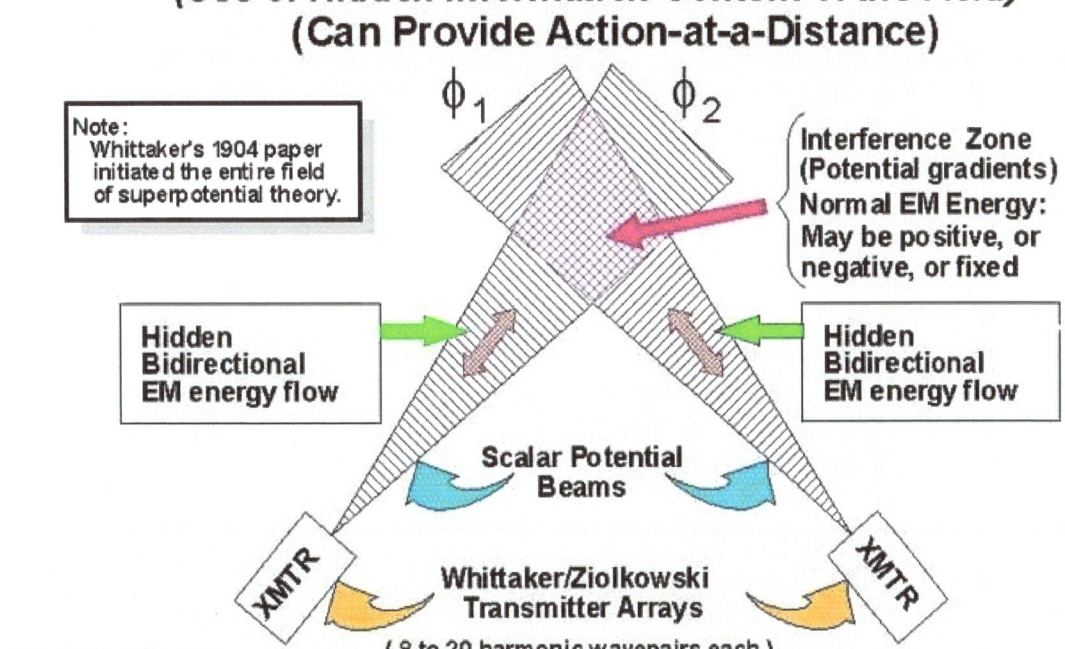

Figure 5. Scalar potential interferometry (between the two sets of bidirectional longitudinal EM wavepair functions) produces all EM force fields and waves.

effects or the symptoms of chemical or biological poisoning; cause a disease epidemic by imprinting the disease 'signature' right into the cellular structure; paralyze and or/kill everyone instantaneously in a 50-mile radius without any detectable warning. These weapons do this by crossing two or more beams with each other and any target can be aimed at, even right through to the opposite side of the earth. The Central Intelligence Agency's MK Ultra Subproject 19 researched the electronic control of human behavior. Subproject 119 had the purpose to provide funds for a study to make a critical review of the literature and scientific developments related to the recording, analysis and interpretation of bio-electric signals from the human organism, and **activation of the human behavior by remote means.** The survey encompassed five main areas: techniques of activation of the human organism by remote electronic means, bio-electric sensors, recording, analysis, and standardization of data. In 1973, with the government-wide panic caused by Watergate, the CIA Director Richard Helms ordered all MKUltra files destroyed. Pursuant to this order, most CIA documents regarding the project were destroyed, making a full investigation of MKUltra impossible – this is the same agency that stated that weapons of mass destruction and al Qaeda were in Iraq! A cache of some 20,000 documents survived Helms' purge, as they had been incorrectly stored in a financial records building and were discovered following a FOIA request in 1977. These documents were fully investigated during the Senate Hearings of 1977.

Scalar potential interferometry – how does it work?

Interference is what happens when two waves carrying energy meet up and overlap. The energy they carry gets mixed up together so, instead of two waves, you get a third wave whose shape and size depends on the patterns of the original two waves. When waves combine like this, the process is called superposition.

Constructive interference means combining two or more waves to get a third wave that's bigger. The new wave has the same wavelength and frequency but more amplitude (higher peaks).

Destructive interference means waves subtracting and canceling out. The peaks in one wave are canceled by the troughs in the other.

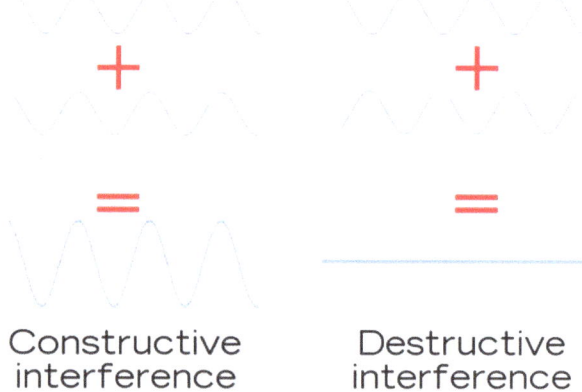

Constructive interference

Destructive interference

www.explainthatstuff.com

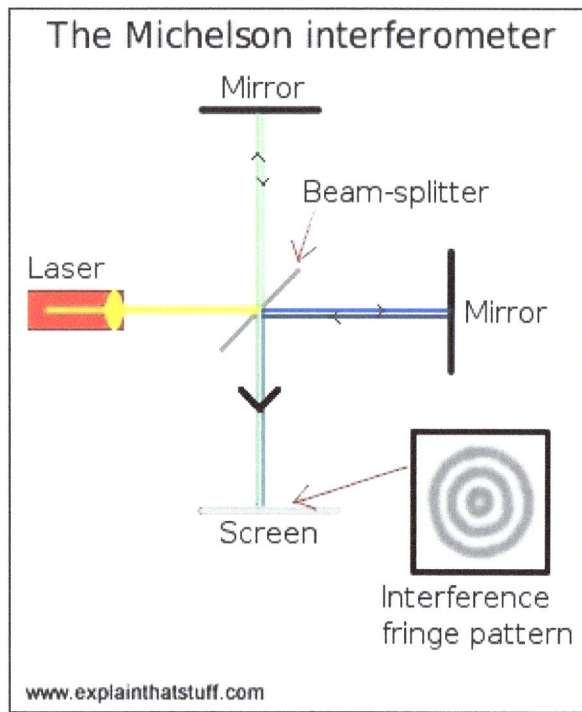

The Michelson interferometer

Mirror

Beam-splitter

Laser

Mirror

Screen

Interference fringe pattern

www.explainthatstuff.com

An interferometer is a really precise scientific instrument designed to measure things with extraordinary accuracy. The basic idea of interferometry involves taking a beam of light (or another type of electromagnetic radiation) and splitting it into two equal halves using what's called a beam-splitter (also called a half-transparent mirror or half-mirror). This is simply a piece of glass whose surface is very thinly coated with silver. If you shine light at it, half the light passes straight through and half of it reflects back—so the beam-splitter is like a cross between an ordinary piece of glass and a mirror. One of the beams (known as the reference beam) shines onto a mirror and from there to a screen, camera, or other detector. The other beam shines at or through something you want to measure, onto a second mirror, back through the beam splitter, and onto the same screen. This second beam travels an extra distance (or in some other slightly different way) to the first beam, so it gets slightly out of step (out of phase).

When the two light beams meet up at the screen, they overlap and interfere, and the phase difference between them creates a pattern of light and dark areas (in other words, a set of interference fringes). The light areas are places where the two beams have added together (constructively) and become brighter; the dark areas are places where the beams have subtracted from one another (destructively). The exact pattern of interference depends on the different way or the extra distance that one of the beams has traveled. By inspecting and measuring the fringes, you can calculate this with great accuracy—and that gives you an exact measurement of whatever it is you're trying to find.

A scalar wave is a purported type of electromagnetic wave that works outside physics as we know it. Scalar Interferometry

The real scientific nomenclature for this weapon is Logitudinal EM Wave Interferometer. Because of it's uneloquent name, mainstream media has dubbed it a "scalar" weapon. The Scalar Interferometry weapon is designed to create electromagnetic fields at a distance. The research started in the 1930's by Nobel Prize winner Tesla. He announced the possibility of a weapon that could destroy hundreds of aircraft hundres of miles away. He came up with this idea half a century before its time. Thus, Tesla was regarded as a mad scientist. He died in 1943 without ever revealing the secret of these great weapons. But he was onto something; research that was not resumed until the late 20th century.

By using longitudinal waves, the energy is triggered to emerge from the local vacuum at the point of the target. Thus a modest triggering signal can cause a huge effect from a distance. This weapon would not have visible light, as most people imagine from movies such as Star Wars. Instead, it uses high-powered microwaves and other sorts of radiation. Though usually deadly to humans, the idea behind these weapons is not to kill, but rather to disrupt electronic systems of enemies. Before the turn of the century, Nikola Tesla had discovered and was utilizing a new type of electric wave. Tesla repeatedly stated his waves were non-Hertzian, and his wireless transmissions did not fall off as the square of the distance. His discovery challenged conventional physics theories and the dominant technologies. It was responsible for the withdrawal of his financial backing, his deliberate isolation from the scientific community, and the gradual removal of his name from the history books.

By 1981 the Soviet Union had discovered and weaponized the Tesla scalar wave effects. Though there are several Tesla devices of note, the most powerful of these frightening weapons was the Tesla Howitzer. It was completed at the Saryshagan missile range and presently considered to be either a high energy laser or a particle beam weapon.

Britain, China, Russia and the United States head up the research in this field, with the United States currently being slightly behind the race.

Seriously flawed path of classical electrodynamics

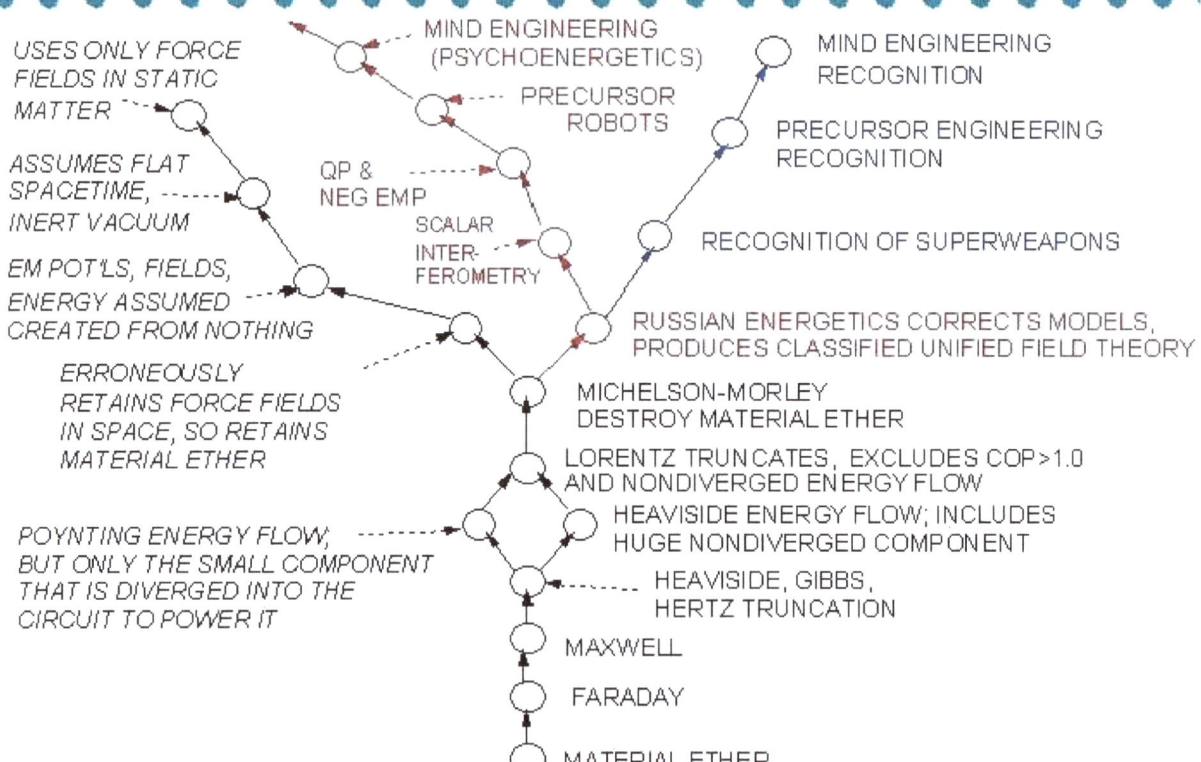

The central concept is that scalar waves restore certain useful aspects of Maxwell's equations. The scientific evolution of classical electrodynamics was "flawed" due to the ommisions in the 19th century by Heaviside, Hertz and Gibbs. Russian energetics scientists corrected the electromagnetic models, and went to develop the *Classified Unified Field Theory*. This led to the development of powerful scalar interferometer superweapons.

According to Thomas Bearden, the Scalar Interferometer is a powerful superweapon that the Soviet Union used for years to modify weather in the rest of the world. It taps the quantum vacuum energy, using a method discovered by T. Henry Moray in the 1920s. The weapon completed at the Russian Saryshagan missile range is a huge Tesla scalar interferometer with four modes of operation. Three of the "modes" are very hard to find any information on at all. However, the one known untilization is called the "Tesla shield." It places a thin, impenetrable hemispherical shell of energy over a large defended area. The 3-dimensional shell is created by interfering two Fourier-expansion, 3-dimensional scalar hemispherical patterns in space so they pair-couple into a dome-like shell of intense, ordinary electromagnetic energy. The air molecules and atoms in the shell are totally ionized and thus highly excited, giving off intense, glowing light. Anything physical which hits the shell receives an enormous discharge of electrical energy and is instantly vaporized.

If several of these hemispherical shells are concentrically stacked, even the gamma radiation and electromagnetic pulse (EMP) from a high altitude nuclear explosion above the stack cannot penetrate all the shells due to repetitive absorption and reradiation, and scattering in the layered plasmas. The altitude of the EMP created by a nuclear blast determines the area where all not-protected electronic devices are disrupted or destroyed.

Hvis flere af disse halvkugleformede skaller er koncentrisk stablet, kan selv den gammastråling og elektromagnetiske impulser (EMP) fra en stor højde atomeksplosion over stablen ikke trænge ind i alle skallerne på grund af gentagne absorption og genudstråling, og spredning i de lagdelte plasmaer. Højden af EMP skabt af en atombombe bestemmer det område, hvor alle ikke-beskyttede elektroniske apparater forstyrres eller ødelægges.

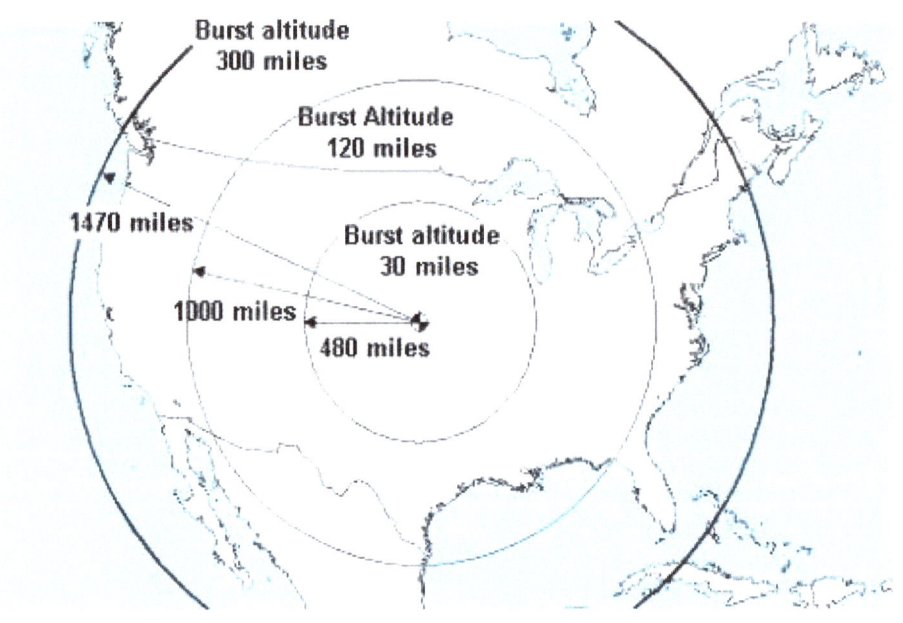

In the 1960s-80s in the USSR there was fulfilled a great number of satellite killers tests. During this warfare there were launched land ballistic missiles, antimissiles, military satellites (including counterweapons) and it made a great impression on the USA. This warfare of the Soviet nuclear force was called "7-hours Nuclear War", it forced America to start creation of antisatellite and anti-missile systems of new generation. "Star Wars" program – the Strategic Defense Initiative proclaimed by Ronald Reagan (1983). Reagan, discussing the Strategic Defense Initiative on a national address, on 23 March 1983, called "upon the scientific community in our country, those who gave us nuclear weapons, to turn their great talents now to the cause of mankind and world peace, to give us the means of rendering those nuclear weapons impotent and obsolete."

Depicted is the Soviet NPO Energia – a military missile station equipped with laser and missile weapons. Based on NPO Energia there were

Боевая ракетная станция конструкции НПО «Энергия»: 1 - базовый блок, включающий в себя агрегатный и приборно-топливный отсеки; 2 - бортовой комплекс вооружения; 3 - самонаводящаяся ракета.

developed two weapon systems: 17F19 "Skif" – applying laser weapons and 17F111 "Kaskad" – a system applying missile weapons.

In 1979 the Soviet Union added a list of some new types of potential weapons of mass destruction:

1) Radiological weapons (using radioactive materials) which could produce harmful effects similar to those of a nuclear explosion;

2)

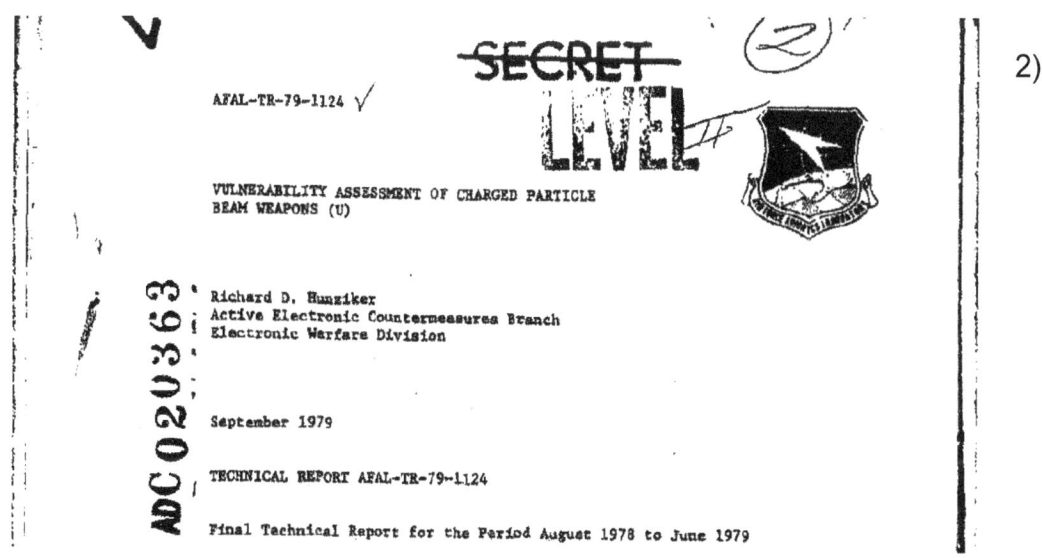

Particle beam weapons, based on charged or neutral particles, to affect biological targets;

3) Infrasonic acoustic radiation weapons;

4) Electromagnetic weapons operating at certain radio-frequency radiations which could have injurious effects on human organs.

In response to the Soviet developments, the US and other Western nations delayed. Directed-energy weapons take the form of lasers, high-powered microwaves and particle beams. Their adoption for ground, air, sea, and space warfare depends not only on using the electromagnetic spectrum, but also upon favorable political and budgetary support.

In an article entitled *"Non-Lethal Weapons May Violate Treaties"*, the author notes that the Certain Conventional Weapons Convention covers many of the non-conventional weapons; "those that utilize infrasound or electromagnetic energy (including lasers, microwave or radiofrequency radiation, or visible light pulsed at brainwave frequency) for their effects."

"Harlan Girard, Managing Director of the International Committee Against Offensive Microwave Weapons, told me he believes the strategy behind the government's recent push for less-than-lethal weapons is a subterfuge. The ones that are now getting all the publicity are put up for scrutiny to get the public's approval. The electromagnetic mind-altering technologies are not mentioned, but would be brought in later under the umbrella of less-than- lethal weapons. These weapons were recently transferred from the Department of Defense over to the Department of Justice. Why? **Because there are several international treaties that specifically limit or exclude weapons of this nature from being used in international warfare.** In other words, weapons that are barred from use against our country's worst enemies (not withstanding the fact that the US did use this weapon against Iraqi troops!) can now be used against our own citizens by the local police departments against such groups as peaceful protestors of US nuclear policies."
http://www.apfn.org/apfn/electronic.htm

The March 23, 1991 newsbrief, "High-tech Psychological Warfare Arrives in the Middle East", describes a US Psychological Operations (PsyOps) tactic directed against Iraqi troops in Kuwait during Operation Desert Storm. The manoeuvre consisted of a system in which subliminal mind-altering technology was carried on standard radio- frequency broadcasts. The March 26, 1991 newsbrief states that among the standard military planning groups in the centre of US war planning operations at Riyadh was "an unbelievable and highly classified PsyOps program utilising 'silent sound' techniques". The opportunity to use this method occurred when Saddam Hussein's military command-and-control system was destroyed.

The Iraqi troops were then forced to use commercial FM radio stations to carry encoded commands, which were broadcast on the 100 MHz frequency. The US PsyOps team set up its own portable FM transmitter, utilising the same frequency, in the deserted city of Al Khafji. This US transmitter overpowered the local Iraqi station. Along with patriotic and religious music, PsyOps transmitted "vague, confusing and contradictory military orders and information."

http://www.abbaswatchman.com/PAGE%2025%20MILITARY%20USE%20OF%20MIND%20CONTROL%20WEAPONS.htm

Subliminally, a much more powerful technology was at work: a sophisticated electronic system to speak directly to the mind of the listener, to alter and entrain his brainwaves, to manipulate his brain's electroencephalographic (EEG) patterns and artificially implant negative emotional states - feelings of fear, anxiety, despair and hopelessness. **This subliminal system doesn't just tell a person to feel an emotion, it makes them feel it, it implants that emotion in their minds.** The application of electromagnetics as a mind-altering mechanism is based on a subliminal carrier technology: the Silent Sound Spread Spectrum (SSSS), sometimes called "S-quad" or "Squad". It was developed by Dr Oliver Lowery of Norcross, Georgia, and is described in US Patent #5,159,703, "Silent Subliminal Presentation System", dated October 27, 1992. The abstract for the patent reads:

"A silent communications system in which nonaural carriers, in the very low or very high audio-frequency range or in the adjacent ultrasonic frequency spectrum are amplitude- or frequency-modulated with the desired intelligence and propagated acoustically or vibrationally, for inducement into the brain, typically through the use of loudspeakers, earphones, or piezoelectric transducers. The modulated carriers may be transmitted directly in real time or may be conveniently recorded and stored on mechanical, magnetic, or optical media for delayed or repeated transmission to the listener."

According to literature by Silent Sounds, Inc., it is now possible, using supercomputers, **to analyse human emotional EEG patterns and replicate them**, then store these "emotion signature clusters" on another computer and, at will, "silently induce and change the emotional state in a human being".

Enligt litteraturen av Silent Sounds, Inc., är det nu möjligt att använda superdatorer , för att analysera mänskliga känslo EEG mönster och replikera dem, sedan lagra dessa "Emotion signatur kluster" på en annan dator och, efter behag, " hemmeligt framkalla och ändra känslomässiga tillstånd i en människa ".

Silent Sounds, Inc. states that it is interested only in positive emotions, but the military is not so constrained. It is obvious that this is a US

Department of Defense project. Edward Tilton, President of Silent Sounds, Inc., describes S-quad in a letter dated December 13, 1996:

"All schematics, however, have been classified by the US Government and we are not allowed to reveal the exact details... ... we make tapes and CDs for the German Government, even the former Soviet Union countries! All with the permission of the US State Department, of course... The system was used throughout Operation Desert Storm (Iraq) quite successfully."

The graphic illustration, "Induced Alpha to Theta Biofeedback Cluster Movement", which accompanies the literature, is labelled #AB 116-394-95 UNCLASSIFIED" and is an output from "the world's most versatile and most sensitive electroencephalograph (EEG) machine". It has a gain capability of 200,000, as compared to other EEG machines in use which have gain capability of approximately 50,000. It is software-driven by the "fastest of computers" using a noise nulling technology similar to that used by nuclear submarines for detecting small objects underwater at extreme range.[6]

The purpose of all this high technology is to plot and display a moving cluster of periodic brainwave signals. The illustration shows an EEG display from a single individual, taken of left and right hemispheres simultaneously. The readout from the two sides of the brain appear to be quite different, but in fact are the same (discounting normal leftright brain variations).

Dr Michael Persinger is a Professor of Psychology and Neuroscience at Laurentian University, Ontario, Canada. Strong electromagnetic fields can affect a person's brain.

"Temporal lobe stimulation," Persinger said, "can evoke the feeling of a presence, disorientation, and perceptual irregularities. It can activate images stored in the subject's memory, including nightmares and monsters that are normally suppressed."

Dr Persinger wrote an article a few years ago, titled "*On the Possibility of Directly Accessing Every Human Brain by Electromagnetic Induction of Fundamental Algorithms*". The abstract reads:

"Contemporary neuroscience suggests the existence of fundamental algorithms by which all sensory transduction is translated into an intrinsic, brain-specific code. Direct stimulation of these codes within the human temporal or limbic cortices by applied electromagnetic patterns may require energy levels which are within the range of both geomagnetic activity and contemporary communication networks. **A process which is coupled to the narrow band of brain temperature could allow all normal human brains to be affected by a subharmonic whose frequency range at about 10 Hz would only vary by 0. 1 Hz.**"

"Within the last two decades a potential has emerged which was improbable, but which is now marginally feasible. **This potential is the technical capability to influence directly the major portion of the approximately six billion brains of the human species,** without mediation through classical sensory modalities, by generating neural information within a physical medium within which all members of the species are immersed.

"The historical emergence of such possibilities, which have ranged from gunpowder to atomic fission, have resulted in major changes in the social evolution that occurred inordinately quickly after the implementation. Reduction of the risk of the inappropriate application of these technologies requires the continued and open discussion of their realistic feasibility and implications within the scientific and public domain."

It doesn't get any plainer than that. And we do not have open discussion because the US Government has totally denied the existence of this technology.

http://www.theforbiddenknowledge.com/hardtruth/mind_control_sounds.htm

A more detailed report on mind control techniques using the electromagnetic spectrum will be the subject of a future book by this author.

Aktiveringen af den menneskelige adfærd ved fjernbetjeningsundersøgelse omfattede fem hovedområder: teknikker til aktivering af den menneskelige organisme ved fjerntliggende elektroniske midler, bio-elektriske sensorer, optagelse, analyse og standardisering af data.

Whittaker showed in 1904 how to turn EM wave energy into electrogravitational potential energy, then how to interfere two such scalar potential waves to recover electromagnetic energy, even at a distance. Whittaker's work is highly significant to modern physics. It produces supersets of quantum mechanics (QM), classical electromagnetics (EM), and general relativity (GR). All three of these extended disciplines are unified with the Whittaker subset, which can be tested and verified by engineering models.

The Soviets in 1985 once threatened the earth itself by activating their scalar weapons with multiple scalar transmitters turned on at once, endangering the survival of the entire planet. Per nuclear physicist Thomas Bearden, they conducted a massive, 'full up' scalar weapon systems and communications strategic exercise. During this exercise, American Frank Golden discovered that the Russians had activated 27 gigantic "'power taps." These were established by resonating the earth electrogravitationally on 54 powerful scalar frequencies (27 pairs where the two are separated from each other by 12 kHz.), transmitted into the earth. The Soviets utilized this to stimulate the earth into forced electrogravitational resonance on all 54 frequencies. Each of the 27 power taps extracted enormous energy from the molten core of the earth itself, and turning it into ordinary electrical power. Each giant tap was capable of powering 4 to 6 of the largest scalar EM howitzers possessed by Russia.

Bearden wrote: "Apparently over 100 giant scalar EM weapons were activated and a large number of command and control transmissions and it lasted several days. By alternating the potentials and loads of each of the two paired transmitters, electrical energy in enormous amounts can be extracted from the earth itself, **fed by the 'giant cathode' that is the earth's molten core**. Scalar EM command and control systems, including high data rate communications with underwater submarines, were also activated on a massive scale. The exercise went on for several days, as power taps were switched in and out, and command and control systems went up and down. Bearden claims not one American intelligence lab, or scientist detected this as they didn't have a detector for scalar EM radiation, and that not one officially believes that the exercise ever happened." However, it was monitored on an advanced, proprietary detection system by Frank Golden for several days and by Bearden for several hours. http://www.angelfire.com/oz/cv/scalarweapons.html

In a conflicting official declassified Navy report (50 Years of Steely Purpose-USS Thresher Remembered. Navsea.navy.mil. 10 April 2013), the disaster was blamed on a one-inch hole in a pipe. The "official" Navy Court of Inquiry concluded that the "*Thresher* had probably suffered the failure of a salt-water piping system joint which relied heavily on silver brazing instead of welding; earlier tests using ultrasound equipment found potential problems with about 14% of the tested brazed joints, most of which were determined not to pose a risk significant enough to require a repair. High-pressure water spraying from a broken pipe joint may have shorted out one of the many electrical panels, causing a shutdown ("scram") of the reactor, with a subsequent loss of propulsion."

Contrary to the Court of Inquiry report, **spurious electromagnetic "splatter" surrounding the immediate vicinity of the targeted area left a signature of intense EM interference with multiple systems and multiple frequencies of the U.S.S. Skylark, surface companion of the Thresher.** This anomalous EM interference was so virulent that it required over 1-1/ 2 hours for the Skylark to transmit an emergency message back to headquarters that the sub had been lost. The death of the Thresher was Khrushchev's first blow. Some electronic systems mysteriously malfunctioned, then later completely recovered spontaneously. This again is clearly a scalar EM weapon signature.
https://www.tldm.org/News8/SovietElectromagneticAttacksOnUnitedStates.htm

The Soviets demonstrated their destructive method of obtaining tremendous electromagnetic energy from the molten core of the earth of the earth, with which to power gigantic strategic scalar Tesla howitzers.

Per Tom Bearden, the Soviet Union had weaponized the Tesla scalar wave effects by 1981. The most powerful of these devastating Tesla weapons -- which Brezhnev undoubtedly was referring to in 1975 when the Soviet side at the SALT talks, suddenly suggested limiting the development of new weapons "more frightening than the mind of man had imagined."

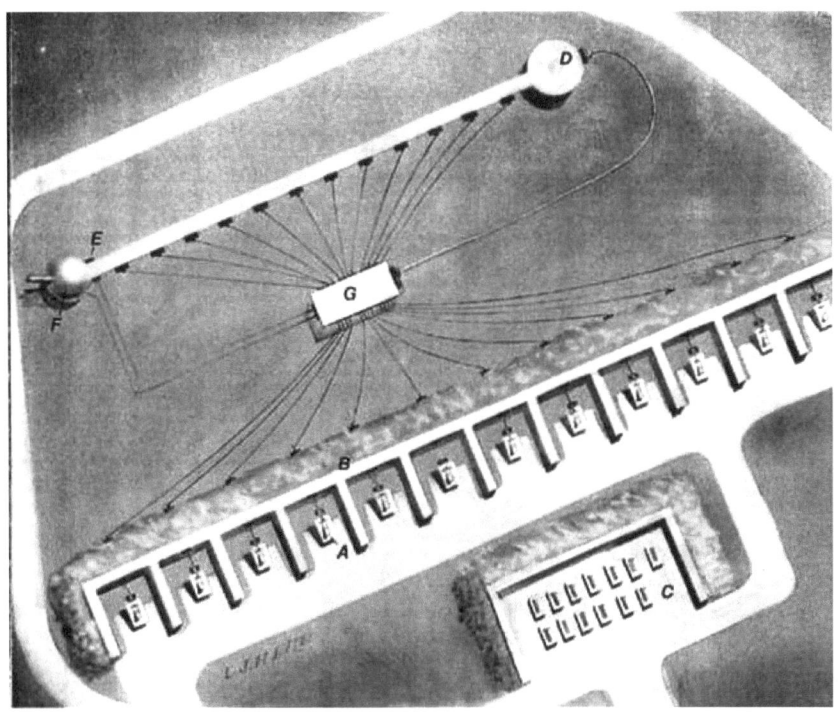

Aviation Week & Space Technology July 28, 1980

As measured by Golden, two signals are continually transmitted into the earth some 12 kilohertz apart, and the earth itself is placed in scalar resonance to the two frequencies. The core of the earth, being a pumped phase conjugate mirror, responds by generating highly amplified phase conjugate replicas, which return and concentrate all their energy at the transmitter/receivers. The receivers receive the enormous beat frequency (12 kHz) energy, using it for power. One then has a kind of scalar interferometer channel in the earth's core in the endothermic mode. Depending upon the biasing, either ordinary EM energy or negative EM energy is captured in the 12-kilohertz band by the interferometer receivers. Varying the negative biasing also provides a throttle to control the amount of energy extracted.

One of these weapons is the Tesla howitzer recently completed at the Saryshagan missile range and presently considered to be either a high-energy laser or a particle beam weapon, (See *Aviation Week & Space Technology*, July 28, 1980, p. 48 for an artist's conception).

The Saryshagan howitzer is a huge Tesla scalar interferometer with four modes of operation. One continuous mode is the Tesla shield, which places a thin, impenetrable hemispherical shell of energy over a large defended area. The 3-dimensional shell is created by interfering two Fourier-expansion, 3-dimensional scalar hemispherical patterns in space so they pair-couple into a dome-like shell of intense, ordinary electromagnetic energy. The air molecules and atoms in the shell are totally ionized and thus highly excited, giving off intense, glowing light. Anything physical which hits the shell receives an enormous discharge of electrical energy and is instantly vaporized -- it goes pfft! like a bug hitting one of the electrical bug killers now so much in vogue.

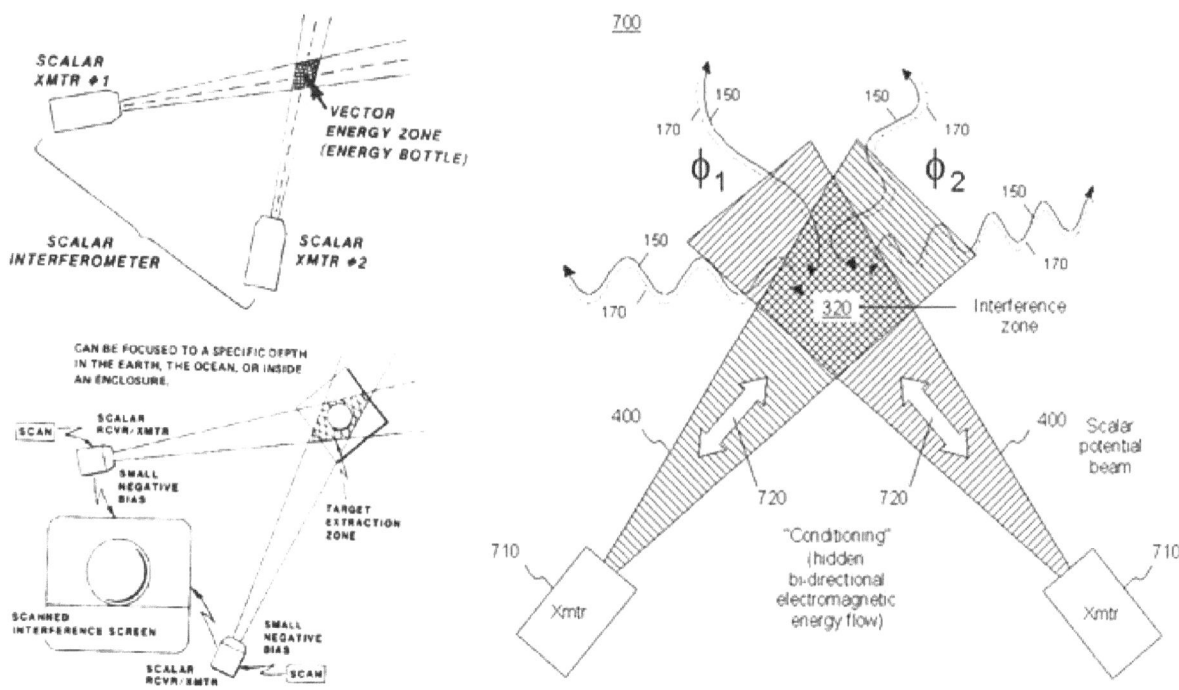

If several of these hemispherical shells are concentrically stacked, even the gamma radiation and EMP from a high altitude nuclear explosion above the stack cannot penetrate all the shells due to repetitive absorption and reradiation, and scattering in the layered plasmas. In the continuous shield mode, the Tesla interferometer is fed by a bank of Moray free energy generators, so that enormous energy is available in the shield.

Aviation Week & Space Technology July 28, 1980

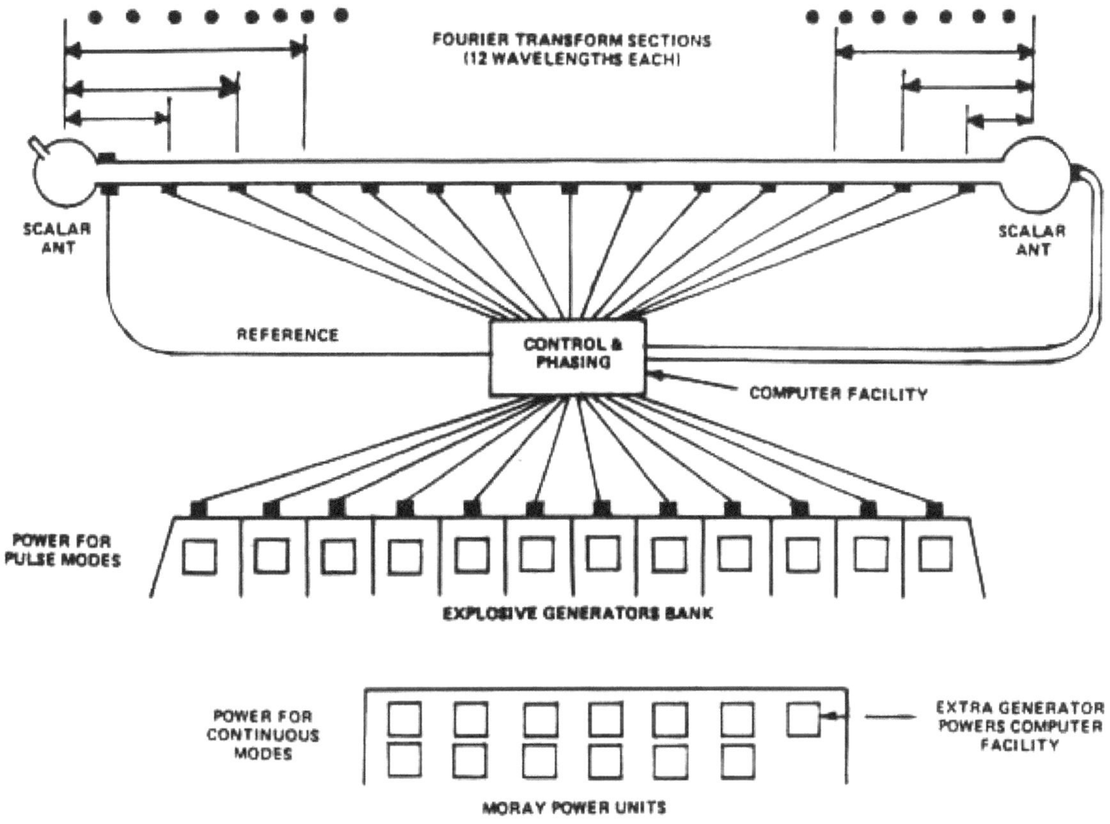

Figure 7. Tesla Weapons at Saryshagan

Hvis flere af disse halvkugleformede skaller er koncentrisk stablet, kan selv den gammastråling og EMP fra en stor højde atomeksplosion over stablen ikke trænge ind i alle skallerne på grund af gentagne absorption og genudstråling, og spredning i de lagdelte plasmaer. Ved den kontinuerlige skjoldtilstand er Tesla interferometeret fødes af en bank af Moray fri energi generatorer, således at enorme energi er tilgængelig i skjoldet.

Den Saryshagan haubits er en enorm Tesla skalar interferometer med fire driftsformer. En kontinuerlig tilstand er Tesla skjold, der placerer en tynd, uigennemtrængelig halvkugleformet skallen af energi over et stort forsvarede område. Det 3-dimensionale skal er lavet ved at blande sig to

Fourier ekspansion, 3-dimensionelle skalar halvkugleformede mønstre i rummet, så de parre-parret til en kuppel-lignende skallen af intens, almindelig elektromagnetisk energi. De luftmolekyler og atom i skallen er fuldstændig ioniseret og kørte derfor meget ophidset, afgiver intense, glødende lys. Alt fysisk Hvilke rammer skallen Modtager ved enorm udladning af elektrisk energi, og er øjeblikkeligt fordampet.

Die Saryschagan Haubitze ist ein großer Tesla skalaren Interferometer mit vier Betriebsarten . Ein Dauerbetrieb ist der Tesla Schild, som legt eine dünne, undurchdringlich hemisphärische Schale Energie über eine große verteidigte Gebiet. Die 3-dimensionale Schale wird durch Interferieren zweier Fourier-Entwicklung, 3-dimensional skalaren hemispherical Muster im Raum, so dass sie paar Paar in eine kuppelartige Schale intensiver, gewöhnliche elektromagnetische Energie geschaffen. Die Luftmoleküle und Atome in der Schale sind vollständig ionisiert und dermed eine sehr aufgeregt, intensive, leuchtende Licht abgibt. Alle physische Objekten, welche die Schale treffen, modtager eine enorme Entlastung von elektrischer Energie, und werden verdampft.

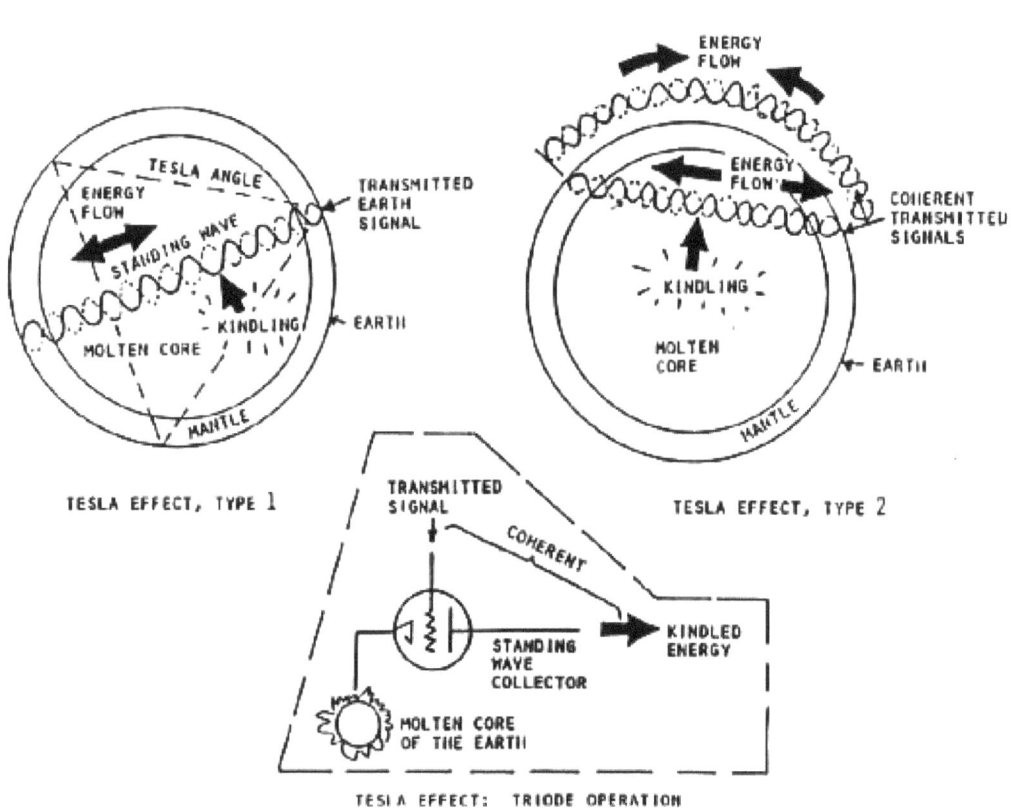

TESLA EFFECT, TYPE 1

TESLA EFFECT, TYPE 2

TESLA EFFECT: TRIODE OPERATION

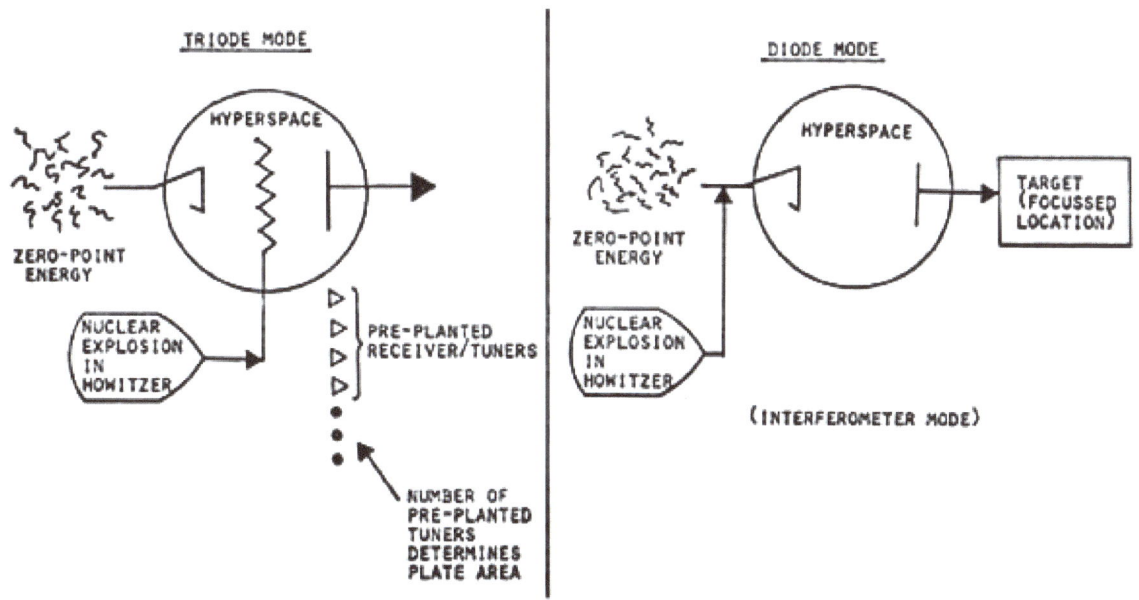

Hyperspace howitzer operation Хиперспаце хаубица операција

http://cheniere.org/books/excalibur/glossary/177.htm

Work on beam weapons also continued in the United States. In 1958 the Defense Advanced Research Projects Agency (DARPA) initiated a top-secret project code-named "Seesaw" at Lawrence Livermore Laboratory to develop a charged-particle beam weapon. More than ten years and twenty-seven million dollars later, the project was abandoned "because of the projected high costs associated with implementation as well as the formidable technical problems associated with propagating a beam through very long ranges in the atmosphere." Scientists associated with the project had no knowledge of Tesla's papers.

In the late 1970s, there was fear that the Soviets may have achieved a technological breakthrough. Some U.S. defense analysts concluded that a large beam weapon facility was under construction near the Sino-Soviet border in Southern Russia.

The American response to this "technological surprise" was the Strategic Defense Initiative announced by President Ronald Reagan in 1983. Teams of government scientists were urged to "turn their great talents now to the cause of mankind and world peace, to give us the means of rendering these nuclear weapons impotent and obsolete."

Lockheed-Martin is already using what is probably a Tesla related invention (they claim it is proprietary), to wirelessly transfer 40W of energy from ground to drones via lasers, enabling them to stay in the air possibly indefinitely. The drone's flight is controlled by an Xbox360 pad.
(EndTheLie / Wired July 13, 2012)

An application of Tesla's dream to positively benefit humanity, is supplying the world with clean, wireless electrical power. An international group out of Russia has launched another fundraiser to go the next step of reproducing Tesla's dream to power the world wirelessly, anywhere, the source of which can be clean. (*PESN* June 8, 2015)

Intel, the world's largest chip manufacturer, has corroborated the experiments of Dr. Konstatin Meyl, by demonstrating a form of wireless energy transfer by lighting a 60-watt bulb from a power source three feet away, in an effect they referred to as WREL (wireless resonant energy link).

The Nevada Lightning Lab has proposed a research facility for generating controlled lightning discharges using a matched set of 12-story Tesla Coil towers, with discharges over 300 feet in length, delivering a peak output of over 18 million volts. Located 35 miles outside Las Vegas, Nevada, this facility will support new industrial and scientific research. The 1:12 scale model twin tower prototypes are completed in San Francisco, CA. (*Slashdot* Dec. 10, 2008).

Free energy devices were proposed by Dr. Holm Hümmler at the 2004 GWUP Conference in Würzburg. "Freie-Energie-Maschinen: Altes Perpetuum Mobile in neuen Schläuchen? Old Perpetual Motion Machines in New Forms?" Sonnabend, 22.5.2004, GWUP Konferenz Würzburg, Dr. rer. nat. Holm Gero Hümmler Dipl.-Phys./Dipl.-Wirt.-Phys.Manager, The Galileo Consulting Group, Ingelheim Kontakt: hummler@web.de

11. DR. KONSTANTIN MEYL – AND THE DISCOVERY OF THE MAGNETIC SCALAR WAVE.

Prof. Konstantin Meyl detected the magnetic scalar wave. Roughly 100 years ago, Nikola Tesla had detected the electric scalar wave, however, this knowledge is suppressed today because it is linked to free energy, which threatens existing energy enterprises. Neutrinos, speeding quicker than light, are an aspect of scalar waves, which can be harvested for free by galaxies, stars, planets, humans etc., giving evidence, that all of them are living entities. In this 3rd interview for White TV, Prof. Meyl gives vital information for physicists, astronomers, astrologers, geologists and free energy researchers on how the universe is build up with neutrinos and how to catch them. In 2006, Meyl could predict an earthquake in connection with a solar eclipse, because he knew about the moon's ability to focus neutrinos. Meyl has described why Kepler's law is not appropriate to describe the movements of stars in a galaxy and why dark matter and string theories are useless nonsense.
https://internationalresearchsociety.wordpress.com/2014/05/16/dr-konstantin-meyl-the-man-they-call-the-new-tesla/

Prof. Meyl's field theory is non-speculative and enables new interpretations of several principles of electrical engineering and quantum physics. This leads to feasible interpretations of experimental observations which to this day have not been possible to explain via existing theories. For example, quantum particle characteristics can be calculated when interpreted as a vortex. The dielectric loss of a capacitor emerges as vortex loss. Likewise, many neutrino experimental results can be explained when the neutrinos are regarded as a vortex. Neutrino power is available as an inexhaustible form of energy due to a remarkable overunity effect. In consideration of environmental sustainability, significant improvements can result by means of this revised theory regarding today's electromagnetic pollution.

Meyl's theory is based on an extension of the Maxwell theory and as such is a special case scenario which does not affect classic physical laws which remain in force. Just as with the spiritual laws, the Commandments, in Matthew 5:17, Jesus assures us that He did not come to destroy the law but to fulfil it. In the physical world of electromagnetics, Tesla, Meyl, and Bearden are "fulfilling" the electromagnetic laws of Maxwell, by incorporating the magnetic flux corrections. In Matthew, 5:19, Jesus exhorts us to not only obey the law but we are to teach it also. We must teach these spiritual and scientific truths and continue to evolve as humans in knowledge and love for one another to survive as a species. In Meyl's enhanced view of the potential vortex, the physical comprehension becomes more objective. Just as Einstein's theory, the Meyl theory explains not only interactions but temperature, which cannot be explained today using conventional theories. Tesla and Meyl have advanced scientific knowledge beyond orthodox scientific knowledge, despite the opposition of the "Scientific Pharisees" of our current age.

"Change is the end result of all true learning."
- Leo Buscaglia

Prof. Meyl has written numerous books. He lectures at the Technical University of Berlin, University of Clausthal and at the University of Applied Sciences Furtwangen. In his end-of-week seminars it is possible to become familiar with the potential vortex, the objective vortex theory and to converse with Prof. Meyl. The current dates are available on his website.

For in-depth information please refer to Prof. Meyl's professional articles and recitations. Available for purchase on this website are books, videos and the remarkable proof of concept transmission-kit. The term "transponder" consists of the terms transmitter and responder, describes thus radio devices which receive incoming signals, to redirect or answer to them. First there were only active transponders, which require a power supply from outside. For some time, passive systems were developed in addition, whose receiver gets the necessary energy at the same time conveyed by the transmitter wirelessly.

After the state of the art several high frequency channels are necessary around the two parts of a transponder system to couple one with another. The transfer of energy from the fundamental unit to the Transceiver takes place with a low frequency, to obtain, as a consequence of the high wavelength, the largest range as possible. The data flow in opposite direction however takes place with high frequencies, which usually already lie in the range of the cellular phone network. Additionally, if data is to be conveyed from the fundamental unit to the transceiver, then a third channel with its own transmitter and receiver is necessary. The enormous expenditure can be reduced to only one channel with substantially larger range. Operation of based on the extended field theory formulated by Meyl.

Two co-workers of Meyl's institute, the *1. Transfer center for scalar wave technology* (www.etzs.de), demonstrated in 2003 at a congress in the technology park of Villingen Schwenningen for the first time publicly, using the ISM frequency of 6.78 MHz, a scalar wave transponder's functionality, consisting of a bi-directional LAN connection to exchange data between two PC's, demonstrating the transfer of energy for the passive interface map over a distance of 30 meters.

In his books, Prof. Dr.-Ing. Konstantin Meyl develops a self-consistent field theory which is used to derive at all known interactions of the potential vortex. Instead of the normally used Maxwell equation, Prof. Meyl chooses Faradays law of induction, as a hypothetical factor and proves that the electric vortex is a part thereof. This potential vortex propagates scalar-like through space and is a longitudinal electric wave whose properties have already been established a century ago by Nicola Tesla.

http://www.meyl.eu/go/index.php?dir=10_Home&page=1&sublevel=0

".. the field equations according to Maxwell do not describe potential vortices in the dielectric such as vortex losses in capacitors. But this was yesterday. Since *Science and Nature* have reported in 2009 the development of magnetic monopoles by the German Helmholtz Society", it is now understood that the 3rd Maxwell equation (divB = 0) only describes a special case. Meyl describes the need to consider "the consequences and the impact to what we learned about electrodynamics." PIERS Proceedings, Stockholm, Sweden, Aug. 12–15, 2013 Consequences of the Extended Field Theory Konstantin Meyl Faculty of Computer and Electrical Engineering, Furtwangen University, Germany

Furthermore, "… the experimental proof of a vortex transmission as a longitudinal wave through air or a vacuum, as accomplished by Tesla already 100 years ago, is neither with Maxwell's field theory nor with the currently used quantum theory explicable or compatible. We are faced with an urgent need for a new field theory."

http://www.k-meyl.de/go/Primaerliteratur/2P9_0930-1-piers-extended_field_theory.pdf

12. DNA-WAVE AND SCALAR WAVE BIOLOGY

Prof. Konstantin Meyl, the Tesla of the 21st Century, has extended his scalar wave research to biology. He detected that <u>our whole body works with scalar waves, that the DNA transports information and energy with a magnetic scalar wave</u>, that the nerves transport scalar waves, that the nodes Ranvier detected indicate a standing, longitudinal wave or scalar wave, and that bio resonance in medical applications and homeopathy use scalar waves.

Dr. Konstatin Meyl has been called "the new Tesla of our time" in an interview for White TV on 7 May 2012. His comprehensive research on potential vortexes, magnetic scalar waves, the unified field theory and the existence of the aether, continues the groundbreaking work that Tesla began more than 100 years ago.

Jag anser att Prof. Dr. Konstantin Meyl är vår tids nye Tesla, supersnillet. I Meyls första intervju för White TV (på engelska) gav han en sammanfattning över hela hans gigantiska forskning som lät honom upptäcka att en virvel består alltid av två, varav den andra var okänd för vetenskapen. Han har döpt den till potentialvirvel. Meyl har återupptäckt skalärvågorna genom att visa hur Tesla upptäckte den elektriska skalärvågen. Meyl har sedan själv upptäckt den magnetiska skalärvågen och det är den som är biologisk relevant. Hela vår kropp arbetar med magnetiska skalärvågor. Meyl formulerade även den enade fältteorin (Einstein misslyckades) och beviset av etern.

In the opinion of the moderator, "I believe that Prof. Dr. Konstantin Meyl is our time, Tesla's new, super-genius. In Meyl's first interview for White TV (in English), he gave a summary of his whole gigantic research that allowed him to discover that a vortex is always made up of two, the second of which was unknown to science. He has named it the potential vortex. Meyl rediscovered scalar waves by demonstrating how Tesla discovered the electric scalar wave. Then Meyl discovered the magnetic scalar wave, and this is the one which is relevant to biology. Our entire human body functions using magnetic scalar waves. Meyl also formulated the unified field theory (where Einstein failed) and the proof of the existence of the ether.

Meyl detected that our whole body works with scalar waves, that the DNA molecule transports information and energy using magnetic scalar waves. DNA, som är byggt som en antenn, transporterar information och energi med hjälp av magnetiska skalärvågor (DNA is constructed like an antenna, transfers information and energy with the help of magnetic scalar waves, according to Meyl.

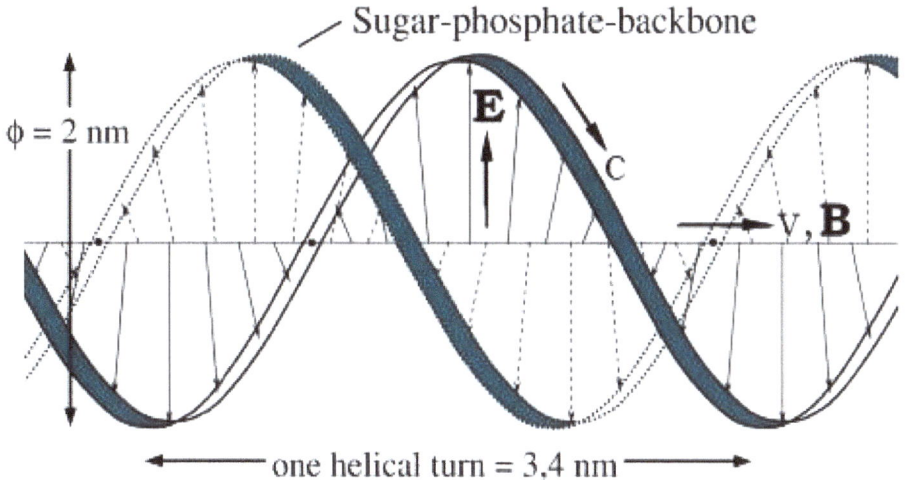

Sugar-phosphate-backbone

$\phi = 2$ nm

E

C

v, **B**

one helical turn = 3,4 nm

Depicted is the distribution of the electric field (E) and of the magnetic flux density (B) in the double helix of a DNA molecule. The propagation speed of a DNA wave is given by v, and is equal to 140,000 km/s. The speed of light is given by c, and is equal to 300,000 km/s, and is equal to the speed of the vortex, as determined by The International CCN Society in 2011.

Nerve cells transport energy and information using scalar waves. Ranvier detected standing longitudinal waves, or scalar waves, at the Nodes of Ranvier, named after him. Bioresonance medicine (Bicom devices) and homeopathy use these waves.

Ranviers noder indikerar en stående våg, en longitudinal våg = skalärvåg och att bioresonans medicinen (Bicom apparaten) och homeopati använder skalärvågor.

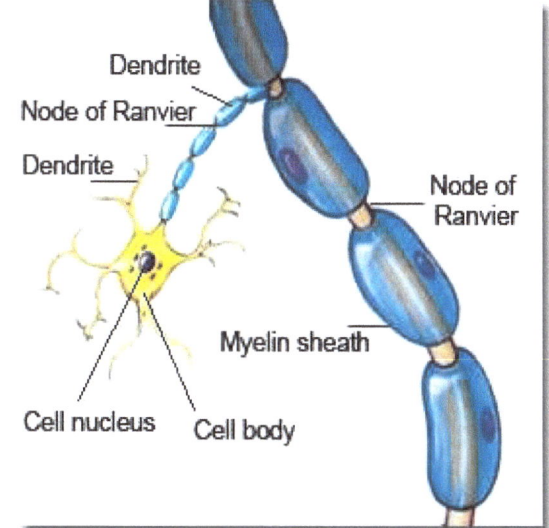

Dendrite

Node of Ranvier

Dendrite

Node of Ranvier

Myelin sheath

Cell nucleus   Cell body

This interview is extremely important for doctors, because Meyl has found the missing links in how the human body functions. But it is also important for teachers to understand, why children have ADHD and difficulty in learning.

Intervjun är mycket viktig för läkare, då Meyl har hittat den felande länken hur vår kropp fungerar. Den är viktig för lärare att förstå varför elever har ADHD och svårigheter att lära sig något.

Meyl also explains what electrosmog is, and how to eliminate it. Victims of tinnitus and mind control can learn a lot, why they are suffering and how they could improve their situation. Victims of tinnitus and mind control can learn a whole lot from Meyl in this interview. He also gives tips about how they can improve their situation.

Meyl förklarar vad elektrosmog är och hur vi kan bli av med det. Tinnitus och mind control (hjärnkontroll) offer kan lära sig mycket av den här intervjun. Meyl ger tips hur de kan förbättra sin situation.

https://internationalresearchsociety.wordpress.com/2014/05/16/dr-konstantin-meyl-the-man-they-call-the-new-tesla/

Additional information on scalar waves is given on the following web sites:

Prof. Meyl New Tesla, Neutrino Free Energy, 100 years Scalar Wave Conspiracy, Our Body works with Scalar Waves, Scalar Weapons on 9/11?, Table Top Defender from QuWave shelter for TI's, Technical proof: Scalar Waves exist; Scalar waves kill cancer cells; The Growing Earth Part 1 and Part 2; Big Bang Big Bluff; Junk DNA no Junk at All!;

While Nikola Tesla was the biggest scientific genius of the 20th Century, the continuation of his revolutionary discoveries is today being advanced by Prof. Dr. Konstantin Meyl, from Villingen, in the Black Forrest of Germany. Meyl has not only organized a revival of Tesla but developed his research to mindboggling levels. He detected that **every vortex has a counterpart, a so-called potential vortex, being able to form matter.** He shows that electrons, protons, neutrinos etc. are double vortexes, without needing a postulate. By that he proves that there are two more types of waves besides the already well-known electromagnetic wave (Hertz 1888). **Tesla (1899) detected the electric scalar wave and Meyl the magnetic scalar wave, giving access to free energy, huge over-unity effects.** The magnetic scalar wave, is integral to neutrinos and consequently exceeds the speed of light, essential to Einstein's theory. Magnetic scalar waves are used in all biological systems.

DNA and Cell Resonance: Magnetic Waves Enable Cell Communication | Abstract

Mary Ann Liebert, Inc. publishers

DNA and Cell Biology

DNA and Cell Resonance: Magnetic Waves Enable Cell Communication

To cite this article:
Konstantin Meyl. DNA and Cell Biology. April 2012, 31(4): 422-426. doi:10.1089/dna.2011.1415.

Published in Volume: 31 Issue 4: April 13, 2012
Online Ahead of Print: October 19, 2011

Die Struktur der DNS – The Structure of the DNS Molecule

„Die Struktur der DNS ist eine sogenannte Doppelhelix. Das Rückgrat der DNS-Kette besteht aus Zuckerphospat. In einer DNS gibt es vier Basen, die jeweils ein Basenpaar bilden, das nur so vorkommen kann. **Adenin (A)-Thymin (T)** und **Cytosin (C)Guanin (G)**.

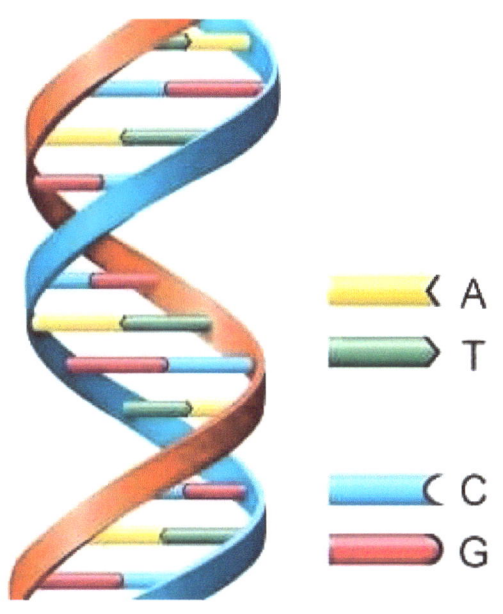

„Bei der Darstellung der Doppelhelix […] ist […] eine Doppelspirale aus 2 Bändern zu sehen, die ähnlich einer Wendeltreppe durch Stufen miteinander verbunden sind. Die Bänder stellen das Rückgrat der [DNS]-Kette dar."
Die Stufen sind die Basenpaare.

A

T

C

G

Die einzigen Unterschiede zu der RNS ist, dass sie keine Doppelhelix ist, sondern nur ein Rückgrat hat und dass anstelle von Thymin die Base **Uracil (U)** steht. Der Ablauf der Basenpaare in der DNS ist der universelle Code für die Erbinformation, d.h. der Ablauf ist bei jedem Lebewesen anders und in genau dieser Reihenfolge ist der genetische Code gespeichert. Die DNS ist nicht nur für die Speicherung, sondern auch für die Weitergabe der Information verantwortlich. Dies geschieht über verschiedene Wege:

Einerseits wird Information bei der Vermehrung der Zellen weitergegeben. Die Zelle und somit auch der Zellkern teilen sich und die DNS wird mitsamt der Information kopiert. Um die DNS kopieren zu können, muss sie entwunden und als Einzelstrang vorhanden sein[1], das heißt „auch die Wasserstoffbrücken zwischen den Basenpaaren müssen gelöst werden."[2] Erst dann kann die DNS vollkommen identisch kopiert werden. Staatliche Fachoberschule Ansbach" (Matthes, *Information in der DNS Evolution oder Schöpfung?*)

Meyl has explained how DNA, performing as antennas, can send information and energy using magnetic waves. He has identified, in *DNA and Cell Resonance: Magnetic Waves Enable Cell Communication*, exactly how electromagnetic waves allow cells to communicate with each other. He also pointed out how the equations of James Clerk Maxwell omitted the critical scalar wave components, which explain the "physical basis of life."

Meyl describes also what electro smog is and how we can avoid this. The cell phone industry, the TV/radio and computer industry and the car manufacturers could revolutionize their products by using scalar waves

[1] Gelhaus, Petra: Gentherapie und Weltanschauung. Ein Überblick über die gen-ethische Diskussion. Univ., Diss--Münster (Westfalen), 2001. Darmstadt: Wiss. Buchges 2006, S. 42.

instead of electromagnetic waves, because they are healthier, use less energy and have one dimension more to transport information. And finally, Meyl discovered unified field theory, which scientists the world over have attempted to do for ages. Einstein, himself, could not succeed in establishing the unified field theory because he based his theories in the false assumption that the speed of light cannot be exceeded, which Tesla discovered from his experiments, and Meyl has now provided conclusive evidence to the contrary.

Konstantin Meyl has identified how electromagnetic waves allow cells to DNA generates a longitudinal wave that propagates in the direction of the magnetic field vector. Longitudinal waves are waves in which the displacement of the medium is in the same direction as, or the opposite direction to, the direction of travel of the wave. After Heaviside's attempts to generalize Maxwell's equations, Heaviside concluded that electromagnetic waves were not to be found as longitudinal waves in "free space" or homogeneous media.

Heaviside, Oliver, "Electromagnetic theory". Appendices: D. On compressional electric or magnetic waves. Chelsea Pub Co; 3rd edition (1971) 082840237X

But Maxwell's equations do lead to the appearance of longitudinal waves under some circumstances, for example, in plasma waves or guided waves. Distinct from the "free-space" waves, such as those studied by Hertz in his UHF experiments, are Zenneck waves.

Corum, K. L., and J. F. Corum, "The Zenneck surface wave", Nikola Tesla, Lightning observations, and stationary waves, Appendix II. 1994.

Computed frequencies from the structure of DNA agree with those of the predicted biophoton radiation. The optimization of efficiency by minimizing the conduction losses leads to the double-helix structure of DNA. The vortex model of the magnetic scalar wave not only covers many observed structures within the nucleus perfectly, but also explains the hyperboloid channels in the matrix when two cells communicate with each other. Potential vortexes are an essential component of a scalar waves, as discovered in 1990. **The basic approach for an extended field theory was confirmed in 2009 with the discovery of magnetic monopoles.**

For the first time, this provides the opportunity to explain the physical basis of life not only from the biological discipline. Nature covers the whole spectrum of known scientific fields of research, and interdisciplinary understanding is required to explain its complex relationships. The characteristics of the potential vortex are significant. With its concentration effect, it provides for miniaturization down to a few nanometers, which allows enormously high information density in the nucleus. With this first introduction of the **magnetic scalar wave, it becomes clear that such a wave is suitable to use genetic code chemically stored in the base pairs of the genes and electrically modulate them, to "piggyback" information from the cell nucleus to another cell.** At the receiving end, the reverse process takes place and the transported information is converted back into a chemical structure. The necessary energy required to power the chemical process is provided by the magnetic scalar wave itself.

Ørsted was aware of moral considerations in science as expressed in his essays Ånden i Naturen (The Spirit in Nature).

With respect to the potentially devastating development of scalar weapons of mass destruction, it is vital for the survivable of the race only to pursue and continue to develop technologies for the betterment of humankind.

 På denne måde udformede Ørsted sig efterhånden den totale verdensopfattelse, han formulerede i skrifter, som for en dels vedkommende samlede i *Aanden i Naturen (1850),* flere afhandlinger om det skønne osv. Det solide naturvidenskabelige grundlag, hvorfra han gik ud, og den omhyggelige tænkning, hvormed han arbejdede sig videre frem, bestandig standsende i tide, gør, at hele hans system endnu den dag i dag i alle væsentlige træk må erkendes for rigtigt, omend en nyere tid end hist og her ville bruge andre udtryk.

Ørsted's Weltanschauung was clearly formulated in his writings on the *Spirit of Nature* (1850), where he highlighted the central importance of beauty and a revelation of the living presence of God.

13. DET AANDELIGE I DET LEGEMLIGE - THE SPIRITUAL IN THE PHYSICAL

I afhandlingen *Det aandelige i det legemlige* viste han, hvorledes det legemlige i sig selv er det forgængelige og omskiftelige, medens det væsentlige i det er de love, hvorefter alle dets virkninger foregå. Da vi ad tankens vej kunne finde en mængde af disse love, kunne de også kaldes fornuftlove eller naturtanker. De er ikke blot tænkte af os; men de have realitet naturen som virkende kræfter. Hele tilværelsen er således et fornuftrige, værk og åbenbaring af den levende alfornuft, Guddommen.

In the thesis, *The spiritual into the physical*, he showed how the physical itself is the ephemeral and changeable, while the essence of it is the laws which govern all its effects. When we by thinking to be able to discover many of these laws, they could also be called common sense laws or natural thoughts. They are not only our thinking; but they are also the acting forces of the reality of nature. The whole of existence is, accordingly, a richly rational work and revelation of the living omniscience of the Godhead.

The first gulp from the glass of natural sciences will turn you into an atheist, but at the bottom of the glass God is waiting for you."
— Werner Heisenberg

Naturlovenes universalitet - The Universality of Natural Laws

I andre afhandlinger i *Aanden i Naturen* påviste Ørsted hvorledes, de sjælelige grundevner væsentlig må være de samme overalt i tilværelsen. En beboer på Jupiter må komme til den samme naturopfattelse som jordbeboeren. Også skønhedslovene og de moralske love må væsentlig være ens i hele universet.

Other papers in the *Spirit of Nature* showed how Ørsted, the fundamental abilities of the soul must be essentially the same everywhere in life. A resident on Jupiter must come to the same conception of nature as an inhabitant of Earth. Also, the laws of beauty and morality must essentially be the same throughout the universe.

Andra teser i *Spirit of Nature* visade Ørsted hur, själens grundläggande förmåga måste vara i huvudsak densamma överallt i livet. En invånare på Jupiter måste komma till samma uppfattning om naturen som en invånare i jorden. Dessutom måste i huvudsak de skönhets och morals lagar vara densamma i hela universum.

Andere Thesen im *Geist der Natur* zeigten, wie Ørsted, die grundlegenden Fähigkeiten der Seele, im wesentlichen, die gleichen überall im Leben sein müssen. Ein Bewohner auf dem Jupiter muss zu der gleichen Vorstellung von der Natur kommen wie ein Bewohner der Erde. Auch müssen die Gesetze der Schönheit und Moral im ganzen Universum dieselben sein.

Om det skønnes væsen handlede adskillige af Ørsteds større og mindre afhandlinger. Han var i grundtrækkene enig med Kant; men hans naturvidenskabelige indsigter gjorde ham det muligt på flere områder at føre teorien videre og anvende den indgående på forskellige konkrete tilfælde, hvorved han gav grundlaget for en virkelig æstetik, der ikke som så mange forfatteres indskrænker sig til blot at være en række lyrisk-subjektive formeninger om det behandlede digterværk, maleri eller deslige, men tværtimod gav en virkelig objektiv analyse af skønhedsfølelsens opståen.

Several of Ørsteds theses dealt with the nature of beauty. In his fundamental ideas, he agreed with Kant, but his scientific insights enable him to develop his theory further and apply it to concrete issues. Unlike many other authors, who confined themselves to subjective evaluations of works of art, Ørsted was able to objectively interpret feelings of beauty.

Også samtlige grundtræk af en etik og en religionsfilosofi vil man finde i hans afhandlinger. Det er vor etiske opgave at få vor vilje og vort liv til at stemme med den evige fornuft. Han tilskriver mennesket en begrænset frihed, men mener ikke, at guddommen af den grund nogensinde behøver at gribe vilkårlig ind i verdensløbet. Denne ytring førte ham til en interessant strid med biskop Jacob Peter Mynster (Aanden i Naturen, 2. del), ligesom han af og til på andre punkter måtte forsvare sine religiøse anskuelser over for samtidens teologi.

Ørsted did not divorce the basic tenets of ethics and his philosophy of religion in his scientific studies. He reminded us that "it is our ethical task to bet our will and our lives in harmony with eternal reason." He ascribes to man a limited freedom, but he did not believe that God did not ever need to intervene arbitrarily into the course of the world.

After a careful investigation of the weaknesses of unbelief and superstition, Ørsted made the following splendid summary:

"...Vantroen bestaar i et Hang til at forkaste *det*, som Mennesker pleier at antage om aandelige Ting, forsaavidt man kun tilegner sig dette ved en umiddelbar, indre Sands, og ikke beviser det ved Tænkningen; den opstaaer i Anledning af de Talrige Tilfælde, hvor Videnskabens Opdagelser gjendrive de Meninger, man uden Undersøgelser havde antaget.

"... Unbelief consists in a tendency to reject that, which people are accustomed to assume about spiritual things, as far as one arrived at faith by an immediate, inner sense, and did not prove it by thought; it is raised because of the numerous instances in which scientific discoveries reject the opinions, one had assumed without investigation."

CONCLUSION

To fail to harness scalar wave technology for the benefit of mankind would be tragic. Our civilization is facing adversity in countless forms. Hunger, disease, pollution, poverty, and war threaten our future and the future of our children. Scalar waves exist, they are useful, and the information to build devices that generate and receive them is available to everyone. To better understand our universe, to ensure the continuation of our civilization, and to honor Nikola Tesla we should do our best to support the open sourcing of this technology.

"The multitude exerts a depressing influence today in Science as it did five centuries ago in the Church. But there always will be individuals of high aims, and endowed by a kind nature with gifts not conforming to the mean. It is to such individuals that the purely scientific work …will prove enticing. While the great multitude will in science only see the modern ladder to position and wealth, these "foolish" individuals will be happy if they obtain an opportunity to devote all their energies to the establishment of the fundamental constants of nature…"

- Gustavus Detlef Hinrichs

The promotion of "individuals of high aims" to leadership positions, and not political and business leaders, who only seek power and wealth, at the expense of humankind, should be the highest priority in assuring the survival of the human race. All institutions, government, business, academic, etc., must support this objective.

We also need to critically examine electromagnetic theory and give credit to Dr. Konstatin Meyl's conclusion that many neutrino experimental results can be explained when the neutrinos are regarded as a vortex. Neutrino power is available as an inexhaustible form of energy due to a remarkable overunity effect. In consideration of environmental sustainability, significant advances can also result by means of Meyl's revised theory regarding electromagnetic pollution.

Critical examination of the research of Bearden and Sweet, and heeding their warnings that "four nations of the world are already embarked on weaponization of scalar EM unified field technology. It is sobering to think that, in addition to having the ability to make our planet a paradise for humankind, we also can make it into a hell. Humankind does not have to allow scalar EM unified field theory to be used to destroy the planet and its citizens. We must not allow scalar technology to repeat the horrible mistakes made with proliferating nuclear energy and encouraging psychopathic warmongers to wage war on God's creations. Heeding John F. Kennedy's words and working to abolish weapons of mass destruction must be our highest calling:

"Every inhabitant of this planet must contemplate the day when this planet may no longer be habitable. Every man, woman and child lives under a nuclear sword of Damocles, hanging by the slenderest of threads, capable of being cut at any moment by accident or miscalculation or by madness. The weapons of war must be abolished before they abolish us."

— John F. Kennedy, Address to the United Nations General Assembly, (25 Sep 1961). On U.S. Department of State website.

Humankind must seize upon the positive aspects of EM technology, and develop and apply this technology for the betterment of all people everywhere. Herbert Hoover once admonished us, regarding the employment of nuclear weapons of mass destruction: "The use of the atomic bomb with its indiscriminate killing of women and children, revolts my soul." Today, we must be just as repulsed and forcibly reject the further development of EM weapons.

Letter (8 Aug 1945) to Colonel John Callan O'Laughlin, publisher of Army and Navy Journal, as quoted in Gar Alperovitz, *The Decision to Use the Atomic Bomb* (1996), 459.

Curriculum in schools and universities must be revised to include the omitted magnetic flux and monopole discoveries, which will permit progress in understand unified gravitational and electromagnetic field theory. Long ago, Albert Einstein said words: "It would of course be a great step forward if we succeeded in combining the gravitational field and the electromagnetic field into a single structure. Only so could the era in theoretical physics inaugurated by Faraday and Clerk Maxwell be brought to a satisfactory close."

As Teilhard de Chardin wrote: "Someday, after we have mastered the winds, the waves, the tides and gravity, we shall harness for God the energies of love. Educating our citizens about the universal concept of love for another, and our Creator, should be our ultimate concern regarding research and implementing scientific knowledge for the benefit of all humanity. Judicious pursuit of scientific knowledge, which improves people's lives, must be our overarching objective. It is necessary to forbid or limit certain scientific pursuits (DNA and GMO manipulation, scalar weaponry development, mind control). Science and the economies of the world should be employed to SERVE humankind, not enslave and diminish people's happiness and freedom.

Scientists, therefore, are responsible for their research, not only intellectually but also morally. This responsibility has become an important issue in many of today's sciences, but especially so in physics, in which the results of quantum mechanics and relativity theory have opened up two very different paths for physicists to pursue. They may lead us—to put it in extreme terms—to the Buddha or to the Bomb, and it is up to each of us to decide which path to take.

— Fritjof Capra, In *The Turning Point: Science, Society, and the Rising Culture* (1983), 87.

Lise Meitner, wife and major contributor to the work of Albert Einstein, was one of the few scientists who courageously opposed the development of the atomic bomb on moral grounds. In response to her invitation in 1943 to work with Otto Robert Frisch and some British scientists at Los Alamos during the Manhattan Project she stated: "I will have nothing to do with a bomb!"

Martinus lived in the country of the great storyteller, Hans Christian Andersen. H. C. Andersen had the ability to make a fairy tale out of reality. One could say the opposite about Martinus; he made the "fairy tale" into reality. He transformed the eternal essence of the religions with their message of limitless love into spiritual science! Only if love for God's creations becomes the highest value of humanity can we limit the use of the powerful technologies discussed in this book to the salvation of the race from its destructive potentialities.

"Then for the second time in the history of the world man will have discovered fire." The authors, Thomas Bearden and Floyd Sweet, fervently believe they have come upon fire for the second time, as allegorized by de Chardin. If so, let us all use the knowledge wisely."

Teilhard de Chardin skrev: »En dag, efter at vi har styr på vinden, bølgerne, tidevandet og tyngdekraften, vi skal udnytte til Gud energierne af kærlighed. Så for anden gang i verdenshistorien mand vil have opdaget brand.. "

Teilhard de Chardin écrivait: «Un jour, après avoir maîtrisé les vents, les vagues, les marées et la gravité, nous aurons recours à Dieu pour les énergies de l'amour, et pour la seconde fois dans l'histoire du monde, l'homme aura découvert le feu. "

Teilhard de Chardin schrieb: "Eines Tages, nachdem wir den Wind, die Wellen, die Gezeiten und die Schwerkraft gemeistert haben, werden wir für Gott die Energien der Liebe nutzen, und dann wird zum zweiten Mal in der Geschichte der Welt der Mensch das Feuer entdeckt haben. "

Teilhard de Chardin skrev: "En dag, efter att vi behärskar vindarna, vågorna, tidvatten och allvar kommer vi att använda energierna av kärlek till Gud, och sedan för andra gången i historien, människan elden har upptäckts."

Russians and Americans do not need to make war. World peace depends on them. Русские и американцы не должны вести войну. Мир во всем мире зависит от них .

President Putin and President Trump understand this. The Democrats want World War Three. Президент Путин и президент Trump понять это. Демократы хотят Третья мировая война.

The Democrats work for the Illuminati, who want war! Демократы работают на иллюминатов, которые хотят войну!

War is very profitable for the Illuminati and Bankers! Война очень выгодно для иллюминатов и банкирами!

I have written three books about the Illuminati. I want people to understand why wars are created.

Я написал три книги о иллюминатов. Я хочу, чтобы люди поняли, почему войны созданы.

We all need to work for world peace, and listen to the Prince of Peace, Jesus! Мы все должны работать для мира во всем мире, и слушать Князя мира, Иисуса!

Bibliography

1. Roberto de Andrade Martins - Romagnosi and Volta's Pile: Early Difficulties in the Interpretation of Voltaic Electricity.

2. Dibner, Bern (1962). *Oersted and the discovery of electromagnetism*, New York, Blaisdell

3. Maxwell, James Clerk (1873). *A treatise on electricity and magnetism Vol I*, Oxford: Clarendon Press.

4. Maxwell, James Clerk (1873). *A treatise on electricity and magnetism Vol II*, Oxford: Clarendon Press.

5. Maxwell, James Clerk (1881). *An Elementary treatise on electricity*, Oxford: Clarendon Press.

6. Maxwell, James Clerk (1890). *The scientific papers of James Clerk Maxwell Vol I*, Dover Publication

7. Maxwell, James Clerk (1890). *The scientific papers of James Clerk Maxwell Vol II*, Cambridge, University Press.

8. Maxwell, James Clerk (1908). *Theory of heat*, Longmans Green Co.

9. Brain, R. M.; et al. (2007). *Hans Christian Ørsted and the Romantic Legacy in Science.* Ideas, Disciplines, Practices. Boston Studies in the Philosophy of Science, *241. Dordrecht. pp. 273–338.* This book owes its origin to the perception of a puzzling paradox. Hans Christian Orsted, the great Danish scientist and philosopher, was one of the founders of modern physics through his experimental discovery in 1820 of the interaction of electricity and magnetism - a key step and model for the further unification of the forces of nature. Followers such as Maxwell and Einstein were, and today searchers worldwide are, enchanted by the hope for a completion of that grand program. In addition to Oersted's discovery of electromagnetism, his work in science included other fields, chiefly high-pressure physics and acoustics. Moreover, he belonged to that fascinating group of researchers who were deeply engaged in the Romantic tradition of the Nature Philosophers, influenced by Immanuel Kant and by religious, literary, and aesthetic currents.

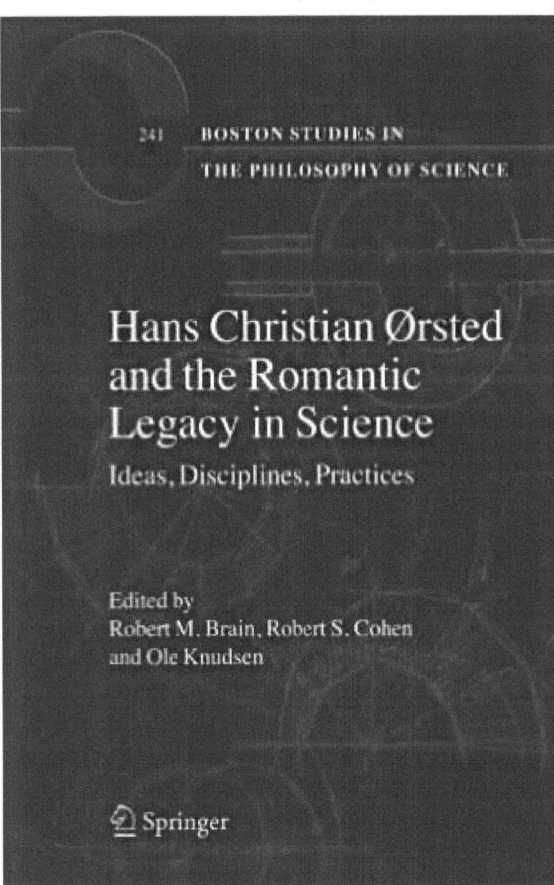

The plain fact is that there are no conclusions. If we must state a conclusion, it would be that many of the former conclusions of the nineteenth-century science on philosophical questions are once again in the melting-pot. — Sir James Jeans, On Free-Will', from Physics and Philosophy (1943), 217. In Franklin Le Van Baumer (ed.), Main Currents of Western Thought (1978), 703.

The scientific and philosophical speculations by Oersted and his circle also quickly stimulated the imagination of other philosophers and scientists, including Gustavus Detlef Hinrichs. Andre-Marie Ampere and Michael Faraday were also inspired by Oersted, whose work launched the transformation of civilization often called the Second Industrial Revolution, based on the invention of motors, generators, and the pervasiveness of electricity in modern life." (Cost: $269.00, on Amazon.com).

10. Henri Poincaré (2006). La valeur de la science. In Anton Bovier, *Statistical Mechanics of Disordered Systems*, p.161.

Les mathématique sont un triple. Elles doivent fournir un instrument pour l'étude de la nature. Mais ce n'est pas tout: elles ont un but philosophique et, j'ose le dire, un but esthétique.
Mathematics has a threefold purpose. It must provide an instrument for the study of nature. But this is not all: it has a philosophical purpose, and, I daresay, an aesthetic purpose.

11. Christensen, D. C. (2013). *Hans Christian Ørsted.* Oxford: Oxford University Press. ISBN 978-0-19-966926-4.

12. Bern Dibner (1962). *Oersted and the discovery of electromagnetism*, New York, Blaisdell.

13. Ole Immanuel Franksen (1981). *H. C. Ørsted – a man of the two cultures*, Strandbergs Forlag, Birkerød, Denmark. (Note: Both the original Latin version and the English translation of his 1820 paper "Experiments on the effect of a current of electricity on the magnetic needle" can be found in this book). H C Ørsted: A Man of the Two Cultures. *Experimenta Circa Effectum Conflictus Electrici in Acum Magneticam*. Hardcover (1981). Price at Amazon.com: Hardcover from $2.41. Also see: *Hans Christian Ørsted, naturfilosof og videnskabsmand*, Kulturcentret Assistens, www.assistens.dk

14. Stringari, Sandro; Robert R. Wilson. *Romagnosi and the discovery of electromagnetism.* Accademia dei Lincei.

15. Krydsfelt, *Ånd og natur i Guldalderen*, redaktion Mogens Bencard, ISBN 87-00-34306-4, side 139.

16. Samlede og efterladte Skrifter af H. C. Ørsted I-IX, 1851-52.

17. H.A.M. Snelders, *Romantiek en 'Naturphilosophie' en de anorganische natuurwetenschappen 1797-1840, Natuurwetenschappen van Renaissance tot Darwin* (Geschiedenis in veelvoud, 18.), Den Haag, 1981, side 168-192, især 178-183.

18. Jelved, Karen & Andrew D. Jackson: *Ørsted, filosofien og fysikken*, Naturens Verden, nr. 2/1999, vol. 82, side 14-19.

19. Kristensen, Marius: "H. C. Ørsted og det danske sprog", Danske Studier 1904, side 49-64.

20. Nielsen, Niels Åge: Sprogrenseren H. C. Ørsted I-II, Aarhus Universitet 1981.

21. Hans Christian Ørsted. *Aanden i Naturen, Fjerde udgave, 1 og 2 Deel*, København 1978, ISBN 87-414-8972-1.

22. Ole Bang, Store Hans Christian, Rhodos, 1986. ISBN 87-7245-149-1.

23. Dan Charly Christensen, *Naturens tankelæser*, 2 bind , Museum Tusculanums Forlag, 2009. ISBN 978-87-635-2524-4 .

24. De Montium Influxu in Valetudinem Hominum, Vitae Genus et Morbos. Dissertatio Inauguralis Medica (Vom Einfluss der Berge auf die Gesundheit der Menschen, auf ihre Lebensweise und ihre Krankheiten). Berlin 1816.

25. *Der Magnetismus nach der allseitigen Beziehung seines Wesens, seiner Erscheinungen, Anwendung und Enträthselung in einer geschichtlichen Entwickelung von allen Zeiten und bei allen Völkern.* Leipzig 1819.

26. *Ueber die nähere Wechselwirkung des Leibes und der Seele, mit anthropologischen Untersuchungen über den Mörder Adolph Moll.* Habicht, Bonn 1825.

27. *Der Magnetismus in seiner geschichtlichen Entwickelung* (Leipzig 1819), ab der 2. Auflage mit dem Titel:

28. Geschichte des thierischen Magnetismus. Bd.: 1 Geschichte der Magie. Leipzig 1844. Neudruck der Ausgabe von 1844, Sändig, Wiesbaden 1966.

29. Historisch-psychologische Untersuchungen über den Ursprung und das Wesen der menschlichen Seele überhaupt, und über die Beseelung des Kindes insbesondere. Bonn 1824, 2. Aufl., Stuttgart 1851.

30. Anthropologische Ansichten zur bessern Kenntnis des Menschen. Bonn 1828.

31. *Der Magnetismus im Verhältnis zur Natur und Religion* (mit einem Anhang über das Tischrücken). Stuttgart 1842, 2. Aufl. 1853.

32. *Was ist die Cholera und wie kann man sich vor ihr am sichersten verwahren?* Nebst Angabe der bewährtesten Heilung derselben. 2. Auflg. Stuttgart 1848.

33. *Der Geist des Menschen in der Natur oder die Psychologie in Uebereinstimmung mit der Naturkunde.* Cotta, Stuttgart 1849.

34. *Anleitung zur Mesmerschen Praxis.* Stuttgart 1852. Neudruck der Ausg. 1852, Kuballe, Osnabrück 1984.

35. *Das Horoskop in der Weltgeschichte.* München 1860.

36. *Das Horoskop in der Weltgeschichte.* Mit d. autobiogr. Fragment: Mein Leben sowie einer symbolischen Figur und einem Brieffaksimile hrsg. u. eingel. von Hermann Haase. Pflüger Verlag, München 1924.

37. Ennemoser, Joseph (1923). *Untersuchungen über den Ursprung und das Wesen der menschlichen Seele - Studies on the Origin and Essence*

of the Human Soul. Mit d. Fragment: Mein Leben. Verlag Die Pforte, Basel 1980. ISBN 3-7725-0184-2. Joseph Ennemoser (15 November 1787 – 19 September 1854) was a South Tyrolean physician and stubborn late proponent of Franz Mesmer's theories of animal magnetism. He became known to English readers through Mary Howitt's translation of his History of Magic (1819, 1844, tr. 1854). After the First Treaty of Paris in 1814, he completed his studies in Berlin and became a supporter of Franz Anton Mesmer and his theory of animal magnetism.

38. Ennemoser Josef. In: Österreichisches Biographisches Lexikon 1815–1950 (ÖBL). Band 1, Verlag der Österreichischen Akademie der Wissenschaften, Wien 1957, S. 254 f. (Direktlinks auf S. 254, S. 255).

39. August Hirsch: Ennemoser, Joseph. In: Allgemeine Deutsche Biographie (ADB). Band 6, Duncker & Humblot, Leipzig 1877, S. 150 f.

40. Jakob Bremm.(1930). *Der Tiroler Joseph Ennemoser: 1787 - 1854; ein Lehrer des tierischen Magnetismus und vergessener Vorkämpfer des entwicklungsgeschichtlichen Denkens in der Medizin.* Fischer, Jena 1930.

41. Karl Wilhelm Schmitz. (1995). *Der Tierische Magnetismus als Teilaspekt der Romantischen Naturphilosophie des frühen 19. Jahrhunderts im Lebenswerk des Tirolers Joseph Ennemoser.* Univ., Diss., Bonn 1995.

42. Werner E. Gerabek: Joseph Ennemoser. In: Werner E. Gerabek, Bernhard D. Haage, Gundolf Keil, Wolfgang Wegner (Hrsg.): Enzyklopädie Medizingeschichte. De Gruyter, Berlin 2005, ISBN 3-11-015714-4, S. 356.

43. Ellen Hastaba, Siegfried de Rachewiltz (Hrsg.). (2009). Für Freiheit, wahrheit und Recht! Joseph Ennemoser und Jakob Philipp Fallmerayer. Tirol von 1809 bis 1848/49. Schlern-Schriften 349, Innsbruck.

44. Monika Fink-Lang: *Der Arzt und Magnetiseur Joseph Ennemoser. Vom Wunder des menschlichen Geistes.* DAMALS Das Magazin für Geschichte 4/ 2010.

45. Siegfried de Rachewiltz (Hrsg.).(2010). *Joseph Ennemoser. Leben und Werk des Freiheitskämpfers, Mediziners und Magnetiseurs (1787 - 1854).* Haymon, Innsbruck. Schriftenreihe historische Quellen zur Kulturgeschichte Tirol Bd 5.

46. Media related to Hans Christian Ørsted at Wikimedia Commons

47. Physics Tree: Hans Christian Ørsted Details

48. Interactive Java Tutorial on Oersted's Compass Experiment National High Magnetic Field Laboratory.

49. *The soul in nature: with supplementary contributions*, London: H. G. Bohn, 1852.

50. Hans Christian Ørsted at Find a Grave.

51. *"Oersted, Hans Christian". Encyclopedia Americana. 1920.*

52. Gitt, Werner. (2002). Hänssler-Verlag, Holzgerlingen.Powerful evidence for the existence of a personal God! Dr. Werner Gitt helps the reader see how the very presence of information reveals a Designer.

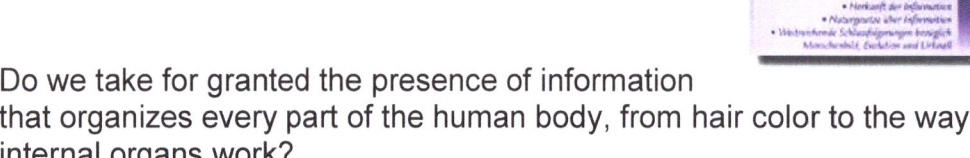

Do we take for granted the presence of information that organizes every part of the human body, from hair color to the way internal organs work?

What is the origin of all our complicated data?

How is it that information in our ordered universe is organized and processed?

Gitt explains the necessity of information - and more importantly, the need for an Organizer and Originator of that information. The huge amount of information present in just a small amount of DNA alone refutes the possibility of a non-intelligent beginning for life. It all points to a Being who not only organizes biological data, but also cares for the creation.

53. Ennemoser, Joseph. (1816). De Montium Influxu in Valetudinem Hominum, Vitae Genus et Morbos. Dissertatio Inauguralis Medica (Vom Einfluss der Berge auf die Gesundheit der Menschen, auf ihre Lebensweise und ihre Krankheiten). Berlin 1816.

DER MAGNETISMUS NACH DER
ALLSEITIGEN BEZIEHUNG SEINES WESENS,
SEINER ERSCHEINUNGEN UND
ENTRÄTSELUNG ......

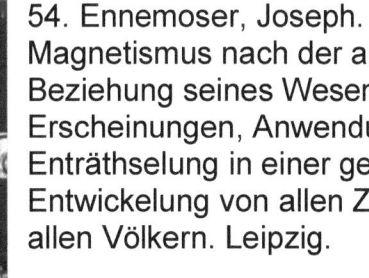

JOSEPH ENNEMOSER

54. Ennemoser, Joseph. (1819). Der Magnetismus nach der allseitigen Beziehung seines Wesens, seiner Erscheinungen, Anwendung und Enträthselung in einer geschichtlichen Entwickelung von allen Zeiten und bei allen Völkern. Leipzig.

Magnetism, per the all-round relation of its essence, its phenomena, application, and discovery in a historical development of all times and among all peoples. Leipzig 1819.

Magnetism, enligt allsidig relation dess väsen, dess fenomen, tillämpning, och upptäckten i en historisk utveckling i alla tider och bland alla folk. Leipzig 1819.

Magnetisme, i henhold til den all-round forhold af sin essens, dens fænomener, ansøgning, og opdagelse i en historisk udvikling af alle tider og blandt alle folkeslag. Leipzig 1819.

Available at Amazon.com, paperback, $55.75.

55. Ennemoser, Joseph. (1825). Ueber die nähere Wechselwirkung des Leibes und der Seele, mit anthropologischen Untersuchungen über den Mörder Adolph Moll. Habicht, Bonn 1825.

56. Ennemoser, Joseph. (1844). *Der Magnetismus in seiner geschichtlichen Entwickelung* (Leipzig 1819), from the 2nd edition with the title Geschichte des thierischen Magnetismus. Bd.: 1 Geschichte der Magie. Leipzig 1844. Facsimile edition, Sändig, Wiesbaden 1966.

57. Ennemoser, Joseph. (1851). *Historisch-psychologische Untersuchungen über den Ursprung und das Wesen der menschlichen Seele überhaupt, und über die Beseelung des Kindes insbesondere.* Bonn 1824, 2. Aufl., Stuttgart 1851.

58. Ennemoser, Joseph. (1828). Anthropologische Ansichten zur bessern Kenntnis des Menschen. Bonn.

59. Ennemoser, Joseph. (1842). *Der Magnetismus im Verhältnis zur Natur und Religion* (mit einem Anhang über das Tischrücken). Stuttgart 1842, 2. Aufl. 1853.

Ennemoser was a leading figure in the then highly fashionable field of 'animal magnetism' (popularised by Mesmer in the later eighteenth century) and hypnosis, and his emphasis on the connection between the mind and physical health foreshadowed Freud's development of psychoanalysis.

The holistic views of the mesmerists incorporated ideas both from natural philosophy and from German Romanticism, and Ennemoser and his contemporaries wrestled with the problem of integrating materialist and mystical viewpoints. In this 1842 publication, Ennemoser analyses the relationship between 'animal magnetism', nature and religion, focusing on phenomena including visions, their physiological and psychological explanations, and the application and effects of 'magnetic' treatments.

Ennemoser var en ledende skikkelse i den daværende meget moderne felt af 'dyr magnetisme "(populariseret af Mesmer i senere attende århundrede) og hypnose, og hans vægt på sammenhængen mellem sind og fysiske sundhed varslet Freuds udvikling af psykoanalysen. De holistiske udsigt over mesmerists indarbejdet ideer både fra naturlige filosofi og fra den tyske romantik, og Ennemoser og hans samtidige kæmpet med problemet med at integrere materialistisk og mystiske synspunkter. I denne 1842 publikation, Ennemoser analyserer forholdet mellem "animalsk magnetisme", natur og religion med fokus på fænomener, herunder visioner, deres fysiologiske og psykologiske forklaringer, samt anvendelse og effekter af 'magnetiske' behandlinger.

Ennemoser var en ledande figur i den då mycket fashionabla området "animalisk magnetism" (populariserades av Mesmer i den senare sjuttonhundratalet) och hypnos, och hans betoning på sambandet mellan sinnet och fysisk hälsa förebådade Freuds utveckling av psykoanalys . Helhets utsikt över mesmerists införlivade idéer från både naturfilosofi och från tyska romantiken, och Ennemoser och hans samtida brottats med problemet att integrera materialist och mystiska synpunkter. I detta 1842 publikation, Ennemoser analyserar förhållandet mellan "djur magnetism", natur och religion, med fokus på fenomen inklusive visioner, deras fysiologiska och psykologiska förklaringar, samt tillämpningen och effekterna av "magnetiska" behandlingar.

Ennemoser war eine führende Persönlichkeit auf dem damals hochmodernen Gebiet des »tierischen Magnetismus« (von Mesmer im späteren achtzehnten Jahrhundert bekannt) und der Hypnose, und seine Betonung auf den Zusammenhang von Geist und Körperlichkeit machte Freuds Entwicklung der Psychoanalyse voraus. Die ganzheitlichen Ansichten der Mesmeristen verkörperten Ideen aus der Naturphilosophie und aus der deutschen Romantik, und Ennemoser und seine Zeitgenossen rangierten mit dem Problem der Integration materialistischer und mystischer Gesichtspunkte. In dieser Publikation von 1842 analysiert Ennemoser die Beziehung zwischen "tierischem Magnetismus", Natur und Religion, die sich auf Phänomene wie Visionen, ihre physiologischen und psychologischen Erklärungen und die Anwendung und Wirkung von "magnetischen" Behandlungen konzentriert.

60. Ennemoser, Joseph. (1849). Ennemoser *Der Geist des Menschen in der Natur oder die Psychologie in Uebereinstimmung mit der Naturkunde*. Cotta, Stuttgart.

61. Ennemoser, Joseph. (1852). Ennemoser *Anleitung zur Mesmerschen Praxis*. Stuttgart 1852. Neudruck der Ausg. Kuballe, Osnabrück 1984.

62. Ennemoser, Joseph. (1860). Ennemoser *Das Horoskop in der Weltgeschichte*. München 1860. Reprinted with an autobiographical fragment: Mein Leben and extra material and commentary by Hermann Haase. Pflüger Verlag, München 1924.

63. Ennemoser, Joseph. (1980). Ennemoser *Untersuchungen über den Ursprung und das Wesen der menschlichen Seele.* Including Mein Leben. Verlag Die Pforte, Basel 1980. ISBN 3-7725-0184-2.

64. Konstatin Meyl (2002). *Wirbelstrukturen als Element der Informationsübertragung in biologischen Systemen*, Kolloquium vom 3.-5.10.2002 Bad Nauheim der BITÄrztegesellschaft, Vortragsband, Seite 77-96.

65. Konstatin Meyl (2003). *Physikalische Grundlagen für die Informationsverarbeitung im Menschen*, Kongressband zu den Festspielgesprächen 2002, (Hg. Simma-Kletschka) Facultas Verlage Wien 2003, Band 27, Wiener Internationale Akademie für Ganzheitsmedizin, ISBN 3-85076-648-9, Seite 38-59.

66. Konstatin Meyl (2004). *Elektromagnetische Umweltverträglichkeit, Teil 3, Skalarwellen und die technische, biologische wie historische Nutzung longitudinaler Wellen und Wirbel*, Umdruck zum informationstechnischen Seminar, (in German), INDEL Verlagsabteilung Villingen-Schwenningen 2002, 3. Aufl. 2004, ISBN 3-9802 542-7-5.

67. Konstatin Meyl (2003). *Scalar Waves, From an extended vortex and field theory to a technical, biological and historical use of longitudinal waves.* Edition belonging to the lecture and seminar „Electromagnetic

Environmental Compatibility ", INDEL Verlagsabt. Villingen-Schwenningen, 1st Ed. 2003, ISBN 3-9802 542-4-0.

68. Konstatin Meyl (2006). *Wireless Tesla Transponder, Field-physical basis for electrically coupled bidirectional far range transponders according to the invention of Nikola Tesla*, SoftCOM 2006, 14th intern. Conference, 29.09.2006, IEEE and Univ. Split, Faculty of Electrical Engineering, ISBN 953-6114-89-5, p. 67-78.

Nikola Tesla (Serbian Cyrillic: Никола Тесла; 10 July 1856 – 7 January 1943) was a Serbian-American inventor, electrical engineer, mechanical engineer, physicist, and futurist who is best known for his contributions to the design of the modern alternating current (AC) electricity supply system.

 Nikola Tesla (Serbian Cyrillic: Никола Тесла; 10. júlí 1856 - 7. Janúar, 1943) var á serbneska-American uppfinningamaður, rafmagnsverkfræðingur vélaverkfræðingur, eðlisfræðingur og futurist sem er best þekktur fyrir framlag sitt til hönnun nútíma riðstraumi (AC) raforkukerfisins.

Nikola Tesla (serbisk kyrillisk: Никола Тесла, 10 juli 1856-7 januar 1943) var en serbisk-amerikanske opfinder, elektroingeniør, maskiningeniør, fysiker, og fremtidsforsker, der er bedst kendt for sine bidrag til udformningen af den moderne vekselstrøm (AC) elsystemet.

Nikola Tesla (serbisch Kyrillisch: Никола Тесла, 10. Juli 1856 - 7. Januar 1943) war ein serbisch-amerikanischer Erfinder, Elektrotechniker, Maschinenbautechniker, Physiker und Futurist, der am besten für seine Beiträge zum Entwurf des modernen Wechselstroms bekannt ist (Wechselstrom).

A gifted researcher and voracious reader, Tesla chanced upon some forgotten volumes of natural science written by Goethe. He had not been aware that Goethe, long before he chose poetry for the vehicle of his scientific themes, had written several magnificent tomes on the natural world. Tesla found to his wonder that Goethe had experienced the very same emotions. …Goethe was aware of the new scientific trend and its implications. <u>The reduction of nature to forces and mechanisms was utterly revolting to Goethe</u>. Now, Tesla found a notable compatriot in his experience. He secured a thorough collection of Goethe's scientific texts and read these to the exclusion of all other philosophies. It was through this window that we may comprehend all of Tesla's scientific methods and later statements. http://www.bibliotecapleyades.net/tesla/esp_tesla_24.htm

For in Tesla we see the quest for communion with nature, one based on the faith that mind, sensation, consciousness, and ordained structure form the world-foundations. The sense-validating *qualitative theme* again appears in Nikola Tesla. Armed with this foundation, he could filter and qualify every other new study with which he was presented.

"The best way to predict the future is to invent it," Peter Drucker said, and that is exactly what Tesla did!

The wireless transmission of electricity, discovered by Tesla, would have prevented the metering of energy by existing power companies and their investors. That is the reason that J.P. Morgan withdrew financing from Tesla when he was building the Wardenclyffe tower on Long Island to supply wireless power to the world.

Den trådløse transmission af elektricitet, opdaget af Tesla, ville have forhindret måling af energi ved eksisterende elselskaber og deres investorer. Det er grunden til, at J. P. Morgan trak finansiering fra Tesla, da han byggede den Wardenclyffe tårn på Long Island for at forsyne trådløs strøm til verden.

The wireless transmission of power and tapping of the energy in the core of the earth, would go a long way to preventing the constant wars in the Middle East, in which many nations are seeking to conquer the oil fields destroying millions of lives and squandering trillions of dollars in fruitless endeavors.

Putting Tesla's ideas to practical use and continuing his development of intercontinental wireless transmission, as his unfinished Wardenclyffe Tower

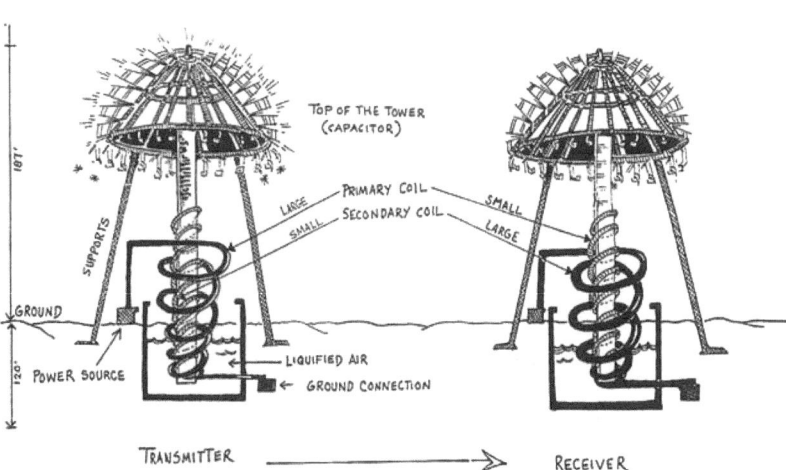

attempted, should be orchestrated by an international control organization, and not be proliferated by nations only desirous of military domination of their neighbors, as is the sad case with nuclear energy. In Tesla's lab, he also conducted a range of experiments with mechanical oscillators/generators, electrical discharge tubes, and early X-ray imaging. He also built a wireless controlled boat, one of the first ever exhibited. Tesla elaborated in a famous article in Century Magazine.

"Stationary waves in the earth mean something more than mere telegraphy without wires to any distance. They will enable us to attain many important specific results impossible otherwise. For instance, by their use we may produce at will, from a sending-station, an electrical effect in any particular region of the globe; we may determine the relative position or course of a moving object such as a vessel at sea, the distance traversed by the same, or its speed; or we may send over the earth a wave of electricity traveling at any rate we desire, from the pace of a turtle up to lightning speed."

An artist's depiction of a solar satellite that could send electric energy by microwaves to a space vessel or planetary surface.

Wireless high power transmission using microwaves is well proven. Experiments in the tens of kilowatts have been performed at Goldstone in California in 1975, and more recently (1997) at Grand Bassin on Reunion Island. These methods achieve distances on the order of a kilometer. Under experimental conditions, microwave conversion efficiency was measured to be around 54%.

In the early 20th century, before the electrical wire grid, Nikola Tesla devoted much effort toward schemes to transport power wirelessly. However, typical configurations (e.g., Tesla coils) required excessively large electric fields. Recent decades have witnessed increased use of autonomous electronic devices (laptops, cell phones, robots, PDAs, etc.). Consequently, interest in wireless power has reemerged. Radiative transfer, which is perfectly suitable for transferring information, poses several difficulties for power transfer applications: The efficiency of power transfer is very low if the radiation is omnidirectional, and unidirectional radiation requires an uninterrupted line of sight and sophisticated tracking mechanisms.

I begyndelsen af det 20. århundrede, før den elektriske ledningsnettet, Nikola Tesla har brugt mange kræfter mod ordninger til at transportere strøm trådløst. Dog typiske konfigurationer (t.e., Tesla spoler), kræves for store elektriske felter. De seneste årtier har været vidne øget brug af autonome elektroniske enheder (bærbare computere, mobiltelefoner, robotter, PDA'er, etc.). Derfor har interesse i trådløs strøm øget. Radiative overførsel, hvilket er perfekt egnet til overførsel af oplysninger, udgør flere vanskeligheder for kraftoverførselssystemersapplikationer: Effektiviteten af kraftoverførsel er meget lav, hvis strålingen er rundstrålende, og ensrettetstrålinger kræver kræver en uafbrudt synslinje og sofistikerede tracking mekanismer.

Wireless power transfer (WPT), wireless energy transmission, or electromagnetic power transfer is the transmission of electrical energy from a power source to an electrical load, such as an electrical power grid or a consuming device, without the use of discrete human-made conductors. Wireless power is a generic term that refers to many different power transmission technologies that use time-varying electric, magnetic, or electromagnetic fields. In wireless power transfer, a wireless transmitter connected to a power source conveys the field energy across an intervening space to one or more receivers, where it is converted back to an electrical current and then used. Wireless transmission is useful to power electrical devices in cases where interconnecting wires are inconvenient, hazardous, or are not possible.

Wireless power techniques mainly fall into two categories, non-radiative and radiative. In near field or non-radiative techniques, power is transferred by magnetic fields using inductive coupling between coils of wire, or by electric fields using capacitive coupling between metal electrodes. Inductive coupling is the most widely used wireless technology; its applications include electric toothbrush chargers, RFID tags, smartcards, and chargers for implantable medical devices like artificial cardiac pacemakers, and inductive powering or charging of electric vehicles like trains or buses. A current focus is to develop wireless systems to charge mobile and handheld computing devices such as cellphones, digital music players and portable computers without being tethered to a wall plug. In far-field or radiative techniques, also called power beaming, power is transferred by beams of electromagnetic radiation, like microwaves or laser beams. These techniques can transport energy longer distances but must be aimed at the receiver. Proposed applications for this type are solar power satellites, and wireless powered drone aircraft.

Japan and China both have national ambitions to begin on-orbit testing of solar power transmission by the 2030s which may accelerate both technical and regulatory progress.

An important issue associated with all wireless power systems is limiting the exposure of people and other living things to potentially injurious electromagnetic fields (see Electromagnetic radiation and health).

Tesla went on to pursue his ideas of wireless lighting and electricity distribution in his high-voltage, high-frequency power experiments in New York and Colorado Springs. He also made early (1893) pronouncements on the possibility of wireless communication with his devices.

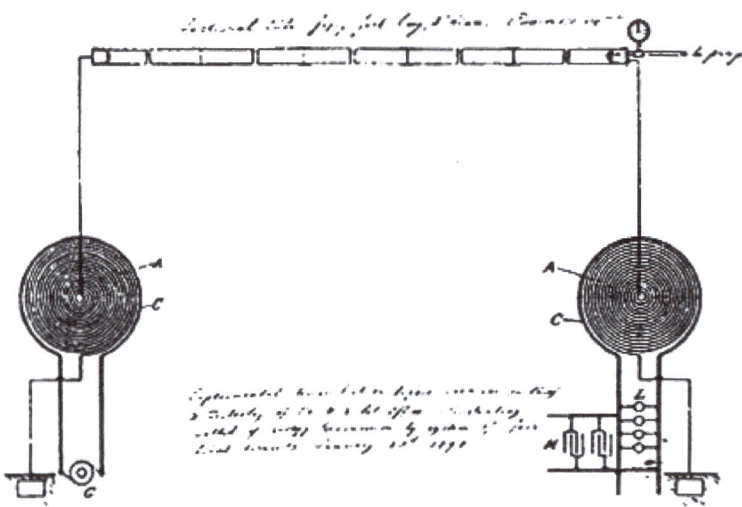

Fig. 1c. Diagram of an apparatus demonstrating transmission of electric power through rarified gas (Tesla's own slide now at the Nikola Tesla Museum, Belgrade)

".......When Nikola Tesla discovered alternating current (AC) electricity, he had great difficulty convincing men of his time to believe in it. Thomas Edison was in favor of direct current (DC) electricity and opposed AC electricity strenuously. Tesla eventually sold his rights to his alternating current patents to George Westinghouse for $1,000,000. After paying off his investors, Tesla spent his remaining funds on his other inventions and **culminated his efforts in a major breakthrough in 1899 at Colorado Springs by transmitting 100 million volts of high-frequency electric power wirelessly over a distance of 26 miles at which he lit up a bank of 200 light bulbs and ran one electric motor! With this souped up version of his Tesla coil, Tesla claimed that only 5% of the transmitted energy was lost in the process.** But broke of funds again, he looked for investors to back his project of broadcasting electric power in almost unlimited amounts to any point on the globe. **The method he would use to produce this wireless power was to employ the earth's own resonance with its specific vibrational frequency to conduct AC electricity via a large electric oscillator.** When J.P. Morgan agreed to underwrite Tesla's project, a strange structure was begun and almost completed near Wardenclyffe in Long Island, N.Y. Looking like a huge lattice-like, wooden oil derrick with a mushroom cap, it had a total height of 200 feet. Then suddenly, Morgan withdrew his support to the project in 1906, and eventually the structure was dynamited and brought down in 1917." http://home.earthlink.net/~drestinblack/wireless.htm

The Nikola Tesla Museum in Belgrade possesses Tesla's own slide which confirms that the experiment described in the patent "System of transmission of electrical energy" was in fact carried out before the Examiner-in-Chief of the U.S. Patent Office.

Tesla was focused in his research for the practical development of a system for wireless transmission of power and a utilization system. Tesla said, in "On electricity", Electrical Review (Jan. 27, 1897):

"In fact, progress in this field has given me fresh hope that I shall see the fulfillment of one of my fondest dreams; namely, the transmission of power from station to station without the employment of any connecting wires."

Tesla's Colorado Springs research was conducted in mid-May 1899. His objectives were to create:

Transmitters of great power.

Individualization and isolating the energy transmission means.

Laws of propagation of currents through the earth and the atmosphere.

Tesla spent more than half his time researching transmitters. Tesla spent less than a quarter of his time researching sensitive receivers and about a tenth of his time measuring the capacity of the vertical antenna. Tesla designed his communications and power broadcast systems based on his model of the earth as a gigantic, spherical capacitor plate, and the ionosphere as the other plate. He discovered the optimal resonant frequencies were 12 Hertz and its harmonics. Also, the "storm" frequency at approximately 500 Kilohertz worked best in his design. Tesla's basic designs of the earth's electrostatic system and his patents are shown below. In accordance with antenna theory, all lengths or circuits are required to be one-quarter wavelength, or some odd multiple of it. This model of the earth's elevated capacitor has two components, the capacity to ground (C_g) and the capacity to the ionosphere (C_i). The values of the resonant frequencies can be determined from the Tesla Equivalent Circuit shown. C_g is the bottom plate to ground, and both plates of this circuit include C_i. L2 and C3 represent the resonant step-down air core coupling

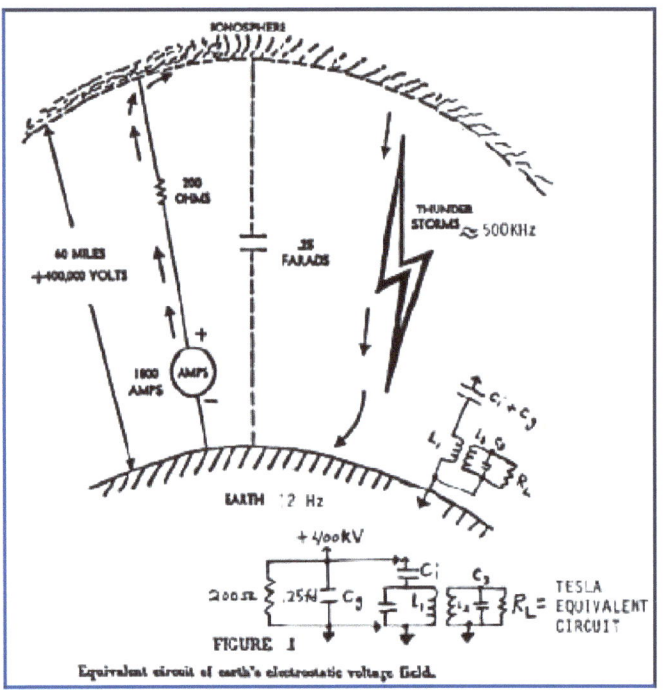

FIGURE 1

Equivalent circuit of earth's electrostatic voltage field.

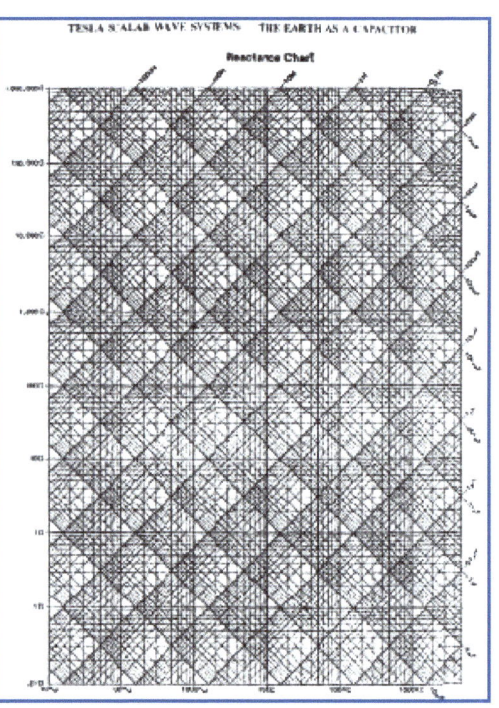

system at the desired frequency. The high voltages Tesla obtained with this design requires extreme caution in duplicating his experiment.

69. Dollard, Eric. (1968). "Theory of Wireless Power." *Wireless Engineer.* Tesla's Wireless Transmission Theory assumed that the oscillating energy surges through the earth to every point on the globe. **In this manner, electric light, heat, and power can be extracted at any point on the earth from a universal, centrally-located station.**

70. Aleksandar Marinčić, Vojin Popović (1999). *Nikola Tesla. Zavod za*

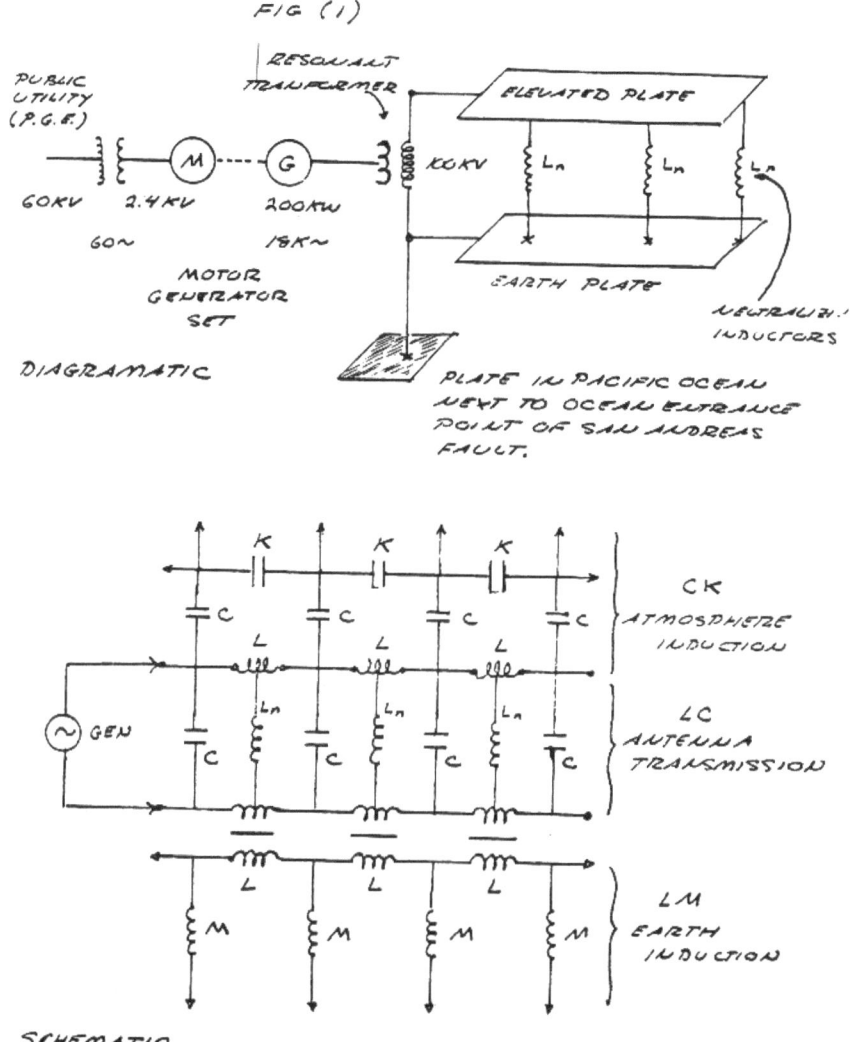

udžbenike i nastavna sredstva.

"From an historical standpoint, it is significant that the genius Nikola Tesla envisaged a worldwide communication system using a huge spark gap transmitter located in Colorado Springs in 1899. A few years later he built a large facility in Long Island that he hoped would transmit signals to the Cornish coast of England. In addition, he proposed to use a modified version of the system to distribute power to all points of the globe".

For experimental verification of his method of wireless power transmission "by conduction through the intervening natural medium", on the global scale Tesla needed still higher voltages and more room. In the Houston Street laboratory, he generated voltages of 2 to 4 MV using a high-frequency transformer with a coil diameter of 244 cm. He these chose Colorado Springs, a plateau about 2000 m above sea level, where he erected a shed large enough to house a high-frequency transformer with a coil diameter of 15 meters!

Although Tesla demonstrated wireless power transmission at Colorado Springs, lighting electric lights mounted outside the building where he had his large experimental coil, he did not scientifically test his theories as stated in the *"The Problem of Increasing Human Energy"* article published in Century Magazine, June 1900. He believed that he discovered the resonance frequency of the earth per W. Bernard Carlson, in *Tesla: Inventor of the Electrical Age*, Princeton University Press - 2013, page 301.

The wireless transmission of electricity, discovered by Tesla, would have prevented the metering of energy by existing power companies and their investors. That is the reason that J.P. Morgan withdrew financing from Tesla when he was building the Wardenclyffe tower on Long Island in order to supply wireless power to the world.

Den trådløse transmission af elektricitet, opdaget af Tesla, ville have forhindret måling af energi ved eksisterende elselskaber og deres investorer. Det er grunden til, at J. P. Morgan trak finansiering fra Tesla, da han byggede den Wardenclyffe tårn på Long Island for at forsyne trådløs strøm til verden.

NASA is also paying attention to scalar waves to some degree. The paper titled, "Advanced Energetics for Aeronautical Applications: Volume II" by David S. Alexander offers a review of scalar wave (longitudinal wave) research. It mentions the work of Konstantin Meyl and others in the field. Interestingly, the paper also discusses how Dr. James Clerk Maxwell's original equations specifically allowed for scalar or longitudinal waves. Only many years later did short sighted scientists who did not care to deal with Maxwell's complex equations based in "quaternion" mathematics arbitrarily discard longitudinal waves. Due to this, many mainstream scientists still doubt their existence, despite the history and the mountain of evidence which proves they are very, very real.

http://pesn.com/2011/03/26/9501797_Teslas_Scalar_Waves_Replicated_by_Steve_Jackson/

Konstantin Meyl has compared superluminal scalar waves to the neutrino, which is also thought by some to travel faster than light. Neutrinos bombard the earth in all directions. Most of them arrive from the central star of our solar system, the sun. Because approximately half of the neutrinos are absorbed as they pass through the Earth, it is thought they may interact with the core and inner layers of our planet.

71. Martinus, Livets bog 2. Martinus describes, from his spiritual perspective, "att vi i stor utsträckning är omgivna av krafter, strålar eller vågor, som vi alls inte kan se och vars existens vi därför tidigare inte anade. Det är med dessa krafters hjälp man är i stånd att skapa de s.k. 'radiovågorna'. Genom dessa 'vågor' har man gjort det till ett känt faktum att hela skådespel, musik och sång kan passera genom vårt hus utan att hindras av väggar, stängda dörrar osv. och uppfylla vårt rum utan att vi märker det, med mindre vi knäpper på en radiomottagare.

I vår atmosfär, i våra omgivningar, genom vår kropp strömmar på samma sätt många energiarter utan att vi direkt kan förnimma eller märka något av dem. Men det innebär naturligtvis inte att de är utan betydelse. Tvärtom medverkar de i högsta grad vid skapandet av grunden för hela vår tillvaro. Hela vår mentalitet eller vårt själsliv är skapat av dessa energier eller ämnen. Dessa ämnen visar sig alltså, som tidigare nämnt, i verkligheten endast vara uttryck för förlängningslinjer av de fysiska ämnena eller materierna, vilka åter [...] endast utgör förtätade medvetenhets- eller själsmaterier." (Livets Bog 2, stycke 294)

In our atmosphere, our surroundings, and through our body many types of energy flow in the same way, which we cannot directly perceive or notice any of them. But that does not mean that they are without significance. On the contrary, they help very much in the creation of the foundation of our entire existence. Our whole mentality or our minds are created by these energies or substances. These substances appear, therefore, as previously mentioned, in reality, only be the expression of the extension lines of the physical elements or matter, which are to be [...] merely condensed consciousness or the materials of the soul." (Livets Bog 2, paragraph 294)

Med "själen" menar Martinus alltså de andliga eller psykiska materier som bygger upp vårt medvetenhetsliv. Fortfarande förefaller dock "själen" ibland ha samma betydelse som "anden". Det gör att innebörden av "ande" och "själ" ändå på någon punkt flyter ihop och kan förstås som två olika sidor av samma verklighet. Men en ännu närmare bestämning av vad man kan mena med "själen" får vi i detta avslutande citat:

"De undermedvetna andliga kropparna, som utgör dags- och nattmedvetandet, är det som i framtidens andliga vetenskap kommer att framträda som det som är 'själen'. Så vi kan säga att det levande väsendet består av ett jag, ett övermedvetande, en själ och en fysisk kropp." (Andligt självmord, Kosmos nr 2 2000)

"The subconscious spiritual bodies, which represents daily and nightly, consciousness, it is like the future spiritual science will emerge as that 'soul'. So, we can say that the living being consists of a self, of a consciousness, a soul and a physical body." (Spiritual suicide, Kosmos No. 2 2000)

72. Konstatin Meyl. (2014). *Dokumentation zur Skalarwellentechnik.* Meyl har sammenlignet superluminal skalare bølger til neutrino, der menes også af nogle at rejse hurtigere end lyset. Neutrinoer bombarderer jorden i alle retninger. De fleste af dem kommer fra den centrale stjerne i vores solsystem, solen. Fordi ca. halvdelen af neutrinoer absorberes når de passerer gennem jorden, menes det at de kan interagere med de centrale og indre lag af vores planet.

Konstantin Meyl hat superluminal Skalarwellen mit der Neutrino verglichen, die auch von einigen angenommen wird, schneller als das Licht zu reisen. Neutrinos bombardieren die Erde in alle Richtungen. Die meisten von ihnen kommen aus dem zentralen Stern unseres Sonnensystems, der Sonne kommen. Da etwa die Hälfte der Neutrinos absorbiert werden, wenn sie durch die Erde passieren, ist es gedacht, sie mit

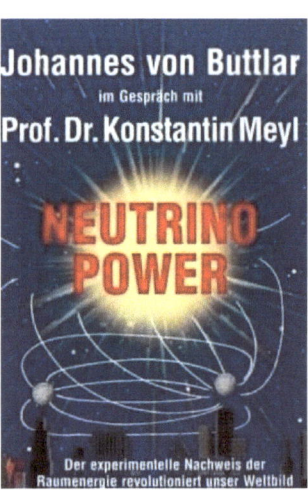

dem Kern und den inneren Schichten unseres Planeten wechselwirken können.

Meyl wirbt für seine Überzeugung, dass die klassische Elektrodynamik auf der Grundlage der Maxwellschen Gleichungen unvollständig sei und durch eine Theorie ersetzt werden müsse, die er selbst begründet hat. Zentrale Begriffe darin sind „Potentialwirbel", „Skalarwelle" oder „Neutrinopower" (auch in Bezug auf die Expansionstheorie der Erde). Er behauptet, damit eine einheitliche Feldtheorie entwickelt zu haben, aus der alle bekannten Wechselwirkungen ableitbar sind. Sie steht allerdings im Widerspruch zu den etablierten physikalischen Theorien der Elektrodynamik, widerspricht auch der Einstein'schen speziellen Relativitätstheorie.

Meyl advocates his conviction that **classical electrodynamics is incomplete based on Maxwell's equations** and must be replaced by a theory which he himself has established. The main concepts are "potential vortex", "scalar wave" or "neutrinopower" (also in relation to the theory of expansion of the earth). **He claims to have developed a uniform field theory**, from which all known interactions can be derived. However, it contradicts the established physical theories of electrodynamics, and contradicts Einstein's special theory of relativity.

Meyl annonserar sin övertygelse om att **klassiska elektrodynamikteorin, baserade på Maxwells ekvationer är ofullständig**, och måste ersättas med en teori som han själv grundade. Nyckelbegrepp i det är "potentiell virvel", "skalär" eller "Neutrino Power" (även i förhållande till expansions teorin om jorden). Han hävdar att han har därför utvecklat en enhetlig fältteori , som alla kända interaktioner kan härledas. Det är dock i motsats till de etablerade fysiska teorier om elektrodynamik och motsäger Einsteins speciella relativitetsteorin .

Meyl annoncerer sin overbevisning om, at **klassisk elektrodynamikteorin, baseret på Maxwell ligninger er ufuldstændig** og skal erstattes af en teori, som han selv grundlagde. Nøglebegreber i det er "potentiale vortex", "skalar" eller "Neutrino Power" (også i forhold til en udvidelse teori af jorden). Han hævder at han har derfor udviklet en forenet feltteori, hvorfra alle kendte interaktioner kan udledes. Men det er i strid med de etablerede fysiske teorier om elektrodynamik, og modsiger Einsteins specielle relativitetsteori.

Meyl annonce sa conviction que **l'électrodynamique classique basé sur les équations de Maxwell sont incomplètes** et doivent être remplacées par une théorie qui lui-même fondé. Les concepts clés en elle sont "vortex potentiel", "scalaire" ou "Neutrino Power" (également en relation avec la théorie de l'expansion de la terre). Il prétend avoir ainsi développé une théorie du champ unifié, à partir de laquelle toutes les interactions connues peuvent être dérivées. Cependant, elle est contraire aux théories physiques établies de l'électrodynamique, contredit la théorie spéciale de relativité d'Einstein.

A popular conspiracy theory holds that Tesla invented a free energy device, but that economic elites has prevented him from publishing it, because it would hurt their investments. Such seizures of research and inventions is entirely legal in the USA since the Invention Secrecy Act of 1951 was passed. The act exists to protect existing infrastructure and industry from overly disruptive or dangerous technology. We must be alert to ruling elites who deny humankind technologies that would improve all lives on earth.

John Jacob Astor
Owner of the Waldorf Astoria hotel

Tesla's intent was to condense the energy trapped between the earth and its upper atmosphere and to turn it into an electric current. He pictured the sun as an immense ball of electricity, positively charged with a potential of some 200 billion volts. The earth, on the other hand, is charged with negative electricity. The tremendous electrical force between these two bodies constituted, at least in part, what he called cosmic energy. It varies from night to day and from season to season, but it is always present.

The positive particles are stopped at the ionosphere and between it and the negative charges in the ground, a distance of 60 miles, there is a large difference of voltage - something on the order of 360,000 volts. With the gases of the atmosphere acting as an insulator between these two opposite stores of electrical charges, the region between the ground and the edge of space traps a great deal of energy. Despite the large size of the planet, it is electrically like a capacitor which keeps positive and negative charges apart by using a non-conducting material as an insulator.

The earth has a charge of 90,000 coulombs. With a potential of 360,000 volts, the earth constitutes a capacitor of .25 farads (farads = coulombs/volts). If the formula for calculating the energy stored in a capacitor ($E = 1/2CV^2$) is applied to the earth, it turns out that the ambient medium contains 1.6×1011 joules or 4.5 megawatt-hours of electrical energy.

To tap this energy storehouse Tesla had to accomplish two things - make a "cold sink" in the ambient energy and devise a way of making the "sink" self-pumping.

The Quantum Energy Generator (QEG) is based on a patent of the famous Serbian-American scientist, inventor and engineer, Nikola Tesla, and was re-designed and modernized by inventor James M. Robitaille. The following is a short description of the QEG by *Fix the World*:

"An average modern household requires 5-10KW of power to operate.

A conventional generator needs 15KW to produce 10KW of power.

To produce these 15KW of power we rely on gas, diesel, propane, coal or other products that can be metered creating profits for the oil industry.

130 years ago, Nikola Tesla invented and patented an energy generator. This is a resonance machine that only needs 1KW of input power to produce 10KW of output power. His patents are now in the public domain.

The *Fix the World Organization* has reproduced Tesla's design with a few modern modifications to generate the same results. Our Quantum Energy Generator (QEG) provides 10KW of power output for less than 1KW input, which it supplies to itself.
Fix the World organisationen har gengivet Teslas design med et par moderne modifikationer til at generere de samme resultater. Vores Quantum Energy Generator (QEG) giver 10KW af effekt for mindre end 1 kW input, som den leverer til sig selv.

We have freely given this technology to the people of the world. We've open sourced a full set of instructions, user manual, schematics and parts list for any engineer to follow and reproduce the same results.

How the QEG works:

Vi har frit givet denne teknologi til de mennesker i verden. Vi har opensourced et komplet sæt af instruktioner, brugermanual, diagrammer og reservedelsliste for enhver ingeniør til at følge og gengive de samme resultater.

First, we use a starting power source, such as an outlet or a crank to power the 1 horsepower motor. This motor spins the rotor in the generator core. The unique oscillator circuitry configuration in the generator core causes resonance to occur. Once the core achieves this resonance it can produce up to 10KW of power, which can then be run through an inverter to power the motor that spins the rotor. You can then unplug the motor from the original power source and the generator will power itself."

Shortly after the announcement and the release of the open source designs, the organization released a video, taken in Morocco, of a working QEG that was switched on in front of a room full of applauding onlookers, apparently showing that the device worked as described. Since then there has been little new information about the QEG, and many people hopeful that this device is the real thing are still wondering if it works or not.

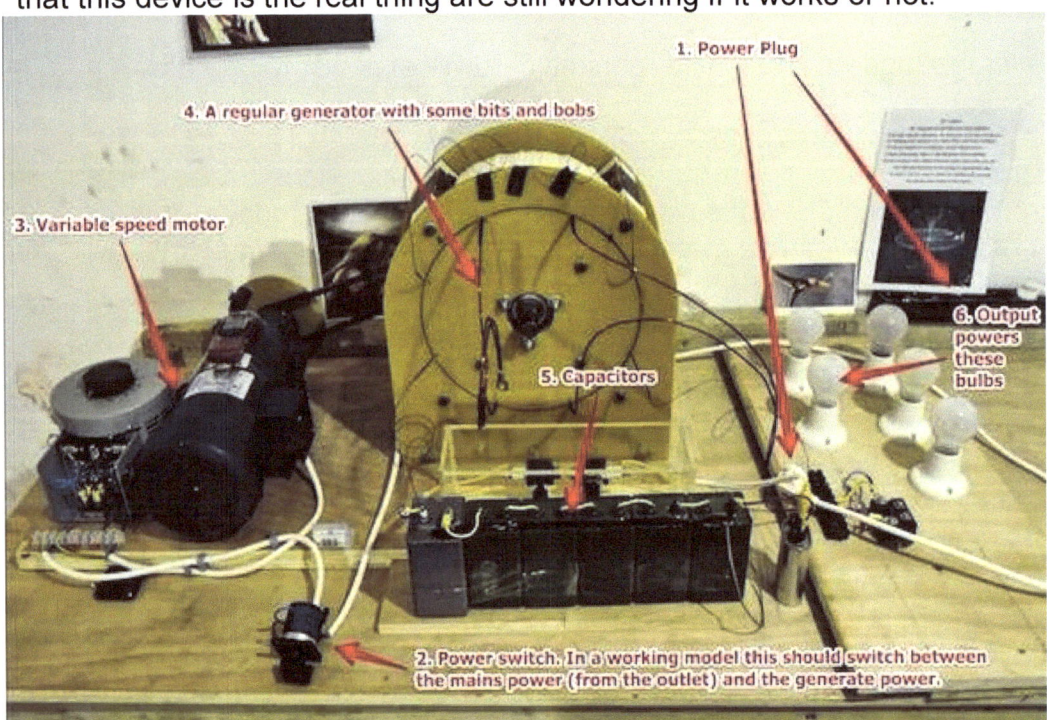

Since the release of the plans for the QEG, and the video of the machine in Morocco, there has been almost no new details about the efficacy of this device. There are very few videos on the web of individual inventors constructing the device, and no one has yet to convince the world that the machine works as advertised. On the contrary, many independent inventors and scientific tinkerers have made various claims that the device is a hoax, and that the demonstrations videos we've seen so far are highly questionable and easy enough to fake. Per recent statements on the *Hope Girl* website, there are currently 60 QEG's being built around the world, and after having experienced some setbacks due to public negativity and betrayal, the organization is "finally getting some of the results we've been looking for!" http://www.wakingtimes.com/2014/12/19/tesla-energy-generator-true-free-energy-device/

A Partial List of Successful Documented EM Over-Unity and Negative Resistor Devices and Processes is listed on Tom Bearden's website. http://cheniere.org/misc/oulist.htm

"There are at least 20 or more legitimate COP>1.0 EM power systems by various inventors and researchers in the U.S. alone"—Tom Bearden

Note: Several working overunity devices can be built from the plans posted on John Bedini's Website. Tom Bearden advises that these devices will work only if they are built exactly as shown, with no deviations or "improvements." The Kawai overunity magnetic motor can also be built directly from the Patent plans, per Bearden.

Medical Applications of Electromagnetics

73. Meyl, Konstatin 2010*). DNA and Cell Resonance – Cellular Communication as Explained by Field Physics Including Magnetic Scalar Waves.* In the water resonance DNA sends a longitudinal wave which propagates within the magnetic field vector. Computed frequencies from DNA structure agree with biophoton radiation frequencies as predicted. Optimization of efficiency is done by minimizing the conduction losses which leads to the double helix structure of DNA. The Meyl vortex model of the magnetic scalar waves not only covers the many observed structures within the nucleus but also introduces the reader to the hyperboloid channels in the matrix as two cells are then found to communicate with each other. Physical results were revealed in1990 which form the theoretical basis of the essential component of a potential vortex scalar wave. An extended field theory approach has been known since 2009 following the discovery of magnetic monopoles. For the first time, magnetic scalar wave theory best explains the physical basis of life not only from the biological discipline of science understanding only. And for the first time this interdisciplinary theory and provides a new understanding of cellular functions that are explained such theory depicting the complex relationships of nature. ...

74. Graille, Jean-Michel. (1984). *Dossier Priore*. Denoël (Editions). (French) ISBN 2207230023

75. Perisse, Eric. (1984). Effects of Electromagnetic Waves and Magnetic Fields on Cancer and Experimental Trypanosomiasis. March 16, 1984. University of Bordeaux.

76. Valone, Thomas (2000). Bioelectromagnetic Healing: A Rationale for its Use. Integrity Research Inst., 2000. ISBN 0-9641070-5-8

77. Bearden, Tom. (1978). U.S. Office of Naval Research Report on the Prioré Machine 16 August 1978.

Dans les années 1960 et 1970, en France, Antoine Prioré a construit et testé des machines de guérison électromagnétique d'une efficacité surprenante. In the 1960's and 1970's, in France, Antoine Prioré built and tested electromagnetic healing machines of startling effectiveness.

Die deutsche Forscherin Andrija Puharich hat während ihren laboratorischen Forschungen bemerkt, dass die Skalar-Energie die Funktion des Immunsystems und des Endokrinsystems verstärkte. The German researcher Andrija Puharich has noted during her laboratory research that scalar energy has enhanced the function of the immune and endocrine systems. http://www.matrixdrops.com/de/ideologie/skalar-energie-eigenschaften-zusammengefasst-von-experten

In hundreds and hundreds of strictly controlled tests with laboratory animals, Prioré's machine cured a wide variety of the most difficult kinds of terminal and fatal diseases known today. In 1974, after a change of local government Prioré lost his government supporters and funding, and subsequent attempts to restore his technology into the public domain were viciously suppressed.

I många strängt kontrollerade tester med försöksdjur, Priorés maskin botade ett brett utbud av de svåraste typer av terminal och dödliga sjukdomar kända idag. I 1974, efter en ändring av kommunernasförvaltningen, Prioré förlorat sina regeringsanhängare och finansiering, och efterföljande försök att återställa sin teknik i det offentliga rummet var brutalt undertryckas. Tom Bearden comments:

"It is my impression that **the French Government did finally recognize how at least part of the Prioré process worked, and weaponized it as longitudinal EM wave interferometers.** In fact, every nuclear weapon on the planet, along with every nuclear powerplant, every nuclear propulsion system, etc. can be dudded in about 10 minutes by one class of these weapons."

- Tom Bearden

"Det är mitt intryck att den franska regeringen slutligen kände igen hur åtminstone en del av Priore processen fungerade , och utvecklas som en längsgående/ longitudinella EM vågsinterferometervapen. I själva verket kan varje kärnvapen på planeten, tillsammans med varje kärnkraftverk, varje kärnframdrivningssystem, etc. kan neutraliseras i ca 10 minuter med en klass av dessa vapen. "

Prioré's machines concretely demonstrated a nearly 100% cure of all kinds of terminal cancers and leukemias, in thousands of rigorous laboratory tests with animals. These results were shown to medical scientists as early as 1960. http://www.cheniere.org/priore/index.html

Prioré's maskiner konkret demonstreret en næsten 100% helbredelse af alle slags terminal kræft og leukæmi, i tusindvis af strenge laboratorieforsøg med dyr. Disse resultater blev vist til medicinske forskere så tidligt som 1960. http://www.cheniere.org/priore/index.html

78. "Oersted". Random House Webster's Unabridged Dictionary.

79. Brian, R.M. & Cohen, R.S. (2007). *Hans Christian Ørsted and the Romantic Legacy in Science*, Boston Studies in the Philosophy of Science, Vol. 241.

80. "Hans Christian Ørsted". Hebrew University of Jerusalem. Retrieved 2009-08-14.

81. Hans Christian Ørsted (1997). Karen Jelved, Andrew D. Jackson, and Ole Knudsen, translators from Danish to English. Selected Scientific Works of Hans Christian Ørsted, ISBN 0-691-04334-5, pp.421-445

82. Martins, Roberto de Andrade, "Resistance to the discovery of electromagnetism: Ørsted and the symmetry of the magnetic field", in: Fabio Bevilacqua & Enrico Giannetto (eds.), Volta and the History of Electricity, Pavia / Milano, Università degli Studi di Pavia / Editore Ulrico Hoepli, 2003, pp. 245-265. (Collana di Storia della Scienza) ISBN 88-203-3284-1

83. Martins, Roberto de Andrade, "Romagnosi and Volta's pile: early difficulties in the interpretation of Voltaic electricity", in Fabio Bevilacqua & Lucio Fregonese (eds.), Nuova Voltiana: Studies on Volta and his Times, Pavia / Milano, Università degli Studi di Pavia / Ulrico Hoepli, 2001, vol. 3, pp. 81-102.

84. Stringari, Sandro; Robert R. Wilson. Romagnosi and the discovery of electromagnetism. Accademia dei Lincei.

85. "Book of Members, 1780–2010: Chapter O" (PDF). American Academy of Arts and Sciences. Retrieved 8 September 2016.

86. "History of DTU". Technical University of Denmark. Retrieved 2009-08-14.

87. National Museum of Denmark. "The Soul in Nature: 1802". Accessed 30 July 2007.

88. Hans Christian, Ørsted (1852). The soul in nature: with supplementary contributions. H. G. Bohn.

89. Heard, John Bickford (1870). *The Tripartite Nature of Man, Spirit, Soul, and Body, Applied to Illustrate and Explain the Doctrines of Original Sin, the New Birth, the Disembodied State, and the Spiritual Body*. T&T Clark.

90. Gustavus Detlef Hinrichs, 1836-1923: 1873-1910. Quelques lettres, en facsimilé, de quelques collègues, membres de l'Institut de France, Académie des sciences, sur le calcul des poids atomiques, sur l'unité de la matière et sur le monument Lavoisier. Adressées: Monsieur le docteur G.-D. Hinrichs ... ([St. Louis, Mo., Woodward & Tiernan Printing Co., 1910]. (page images at HathiTrust)

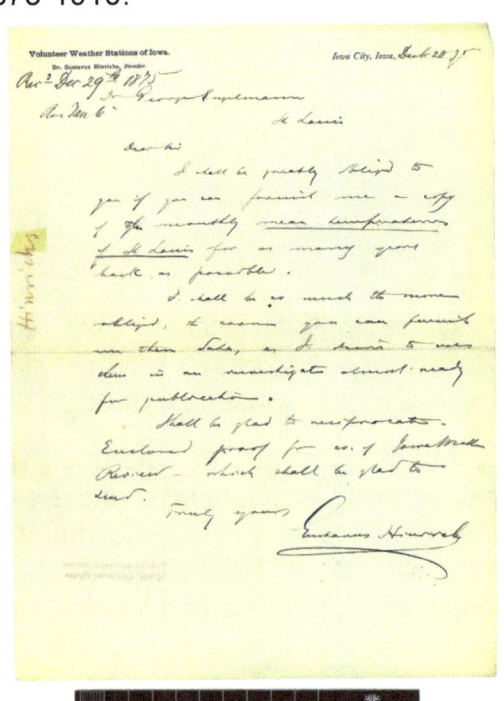

91. Hinrichs, Gustavus Detlef, 1836-1923: The absolute atomic weights of the chemical elements, established upon the analyses of the chemists of the nineteenth century and demonstrating the unity of matter; presented in simple language to the general scientific public ... (St. Louis, C. G. Hinrichs, 1901) (page images at HathiTrust)

92. Hinrichs, Gustavus Detlef, 1836-1923: The Amana meteorites of February 12, 1875, (St. Louis, C. G. Hinrichs, 1905) (page images at HathiTrust)

93. Hinrichs, Gustavus Detlef, 1836-1923: Contributions to molecular science, or atomechanics ... (Iowa-City, 1868-70) (page images at HathiTrust)

94. Hinrichs, Gustavus Detlef, 1836-1923: The elements of physics, demonstrated by the student's own experiments. (Davenport, Ia.: Griggs, Watson & Day; Leipzig, F.A. Brockhaus, 1870) (page images at HathiTrust)

95. Hinrichs, Gustavus Detlef (1867). *Programme der atomechanik*, Iowa-City. (page images at HathiTrust).

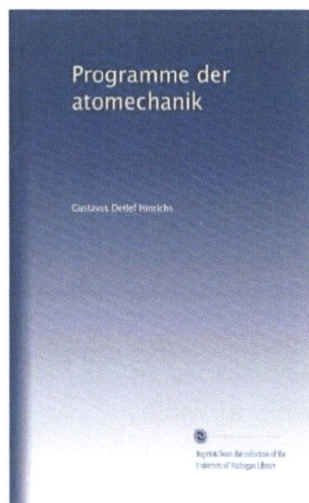

Hinrichs, like Tesla, lamented the view of the masses to only value science in materialistic terms. "The multitude exerts a depressing influence today in Science as it did five centuries ago in the Church. But there always will be individuals of high aims, and endowed by a kind nature with gifts not conforming to the mean. It is to such individuals that the purely scientific work here mapped out will prove enticing. While the great multitude will in science only see the modern ladder to position and wealth, these "foolish" individuals will be happy if they obtain an opportunity to devote all their energies to the establishment of the fundamental constants of nature, the true and exact values of the atomic weights of the elements. To these true chemists, to whom we must look for the great work indicated, I say with Dante (III, x, 22-25):

Or ti riman, Lettor, sovra'l tuo banco, Dietro pensando a cio, che si preliba, S'esser vuol lieto assai prima, che stance, Messo t'ho inanzi: omai per te ti ciba."

Long before Nicola Tesla's revolutionary ideas and patents were misappropriated by commercial and academic thieves, Hinrichs lamented the plagiarizing of his contribution in pages 239 – 240 of his *The Periodic Law*. "A couple of years after the publication of my *Programme der Atomechanik* the so-called Periodic Law was presented to the chemical world by Lothar Meyer and by Mendelejeff. In a period of empiricism, a l'outrance this "law" was promptly adopted and incorporated in text-books and presented in most of the chemical lecture halls throughout the world. Being universally known, it is not necessary for me to give any details concerning the same. Our only object is to show that so far as it was new, it is not true, and so far, as it is true, it was not new, but explicitly contained in my Programme of 1867. One of these discoverers of the Periodic Law, Lothar Meter was well acquainted with my *Atomechanik*, for he himself published a review thereof. The other discoverer, Mendelejeff, shows a little more originality in the handling of the subject, and may possibly have arrived at the result independently, though copies of my work were accessible to him. However, his publication was at least two years later than mine. I am not willing to devote much space to this subject; it is not worth it, and the question is one that the future must and will pass upon. The following, it is hoped, will suffice for the present. The elements are not mere periodic functions of their atomic weight, in any sense of the word. All attempts to arrange the elements in regular, continued periods have absolutely failed. Recurrences are real, but not at regular, equal intervals. The chemical and physical absurdities involved in this periodic law are, by common consent, overlooked to-day; but they are visible to all who will use their eyes — and some chemists will have their eyes open in the future. Furthermore, the very idea is mathematically absurd; a periodic function always implies more than one variable. If the atomic weight be the one compared, say to time, in astronomy— what is the other variable, corresponding to the orbit? Mendelejeff shows that he has no conception of the subject."

The need to employ the scientific method, and incorporate a multidisciplinary approach to advancing scientific knowledge was pointed out by Hinrichs on page 171. "It is interesting to notice that in the case of astronomy, depending for its facts upon observation, it has required twenty-five centuries of careful observation to recognize the existence of this secular variation, while in Chemistry, facts being obtainable at will by experiment, it was sufficient to increase the weight of silver used from one to two hundred and fifty grammes, in order to produce the variation of the atomic weight of Sulphur so strikingly shown in our diagrams representing the experiments of Cooke, Dumas and Stas. If our modern chemists, who insist on being recognized as leaders in our science, had devoted some of their time of study to those branches of the science of Nature which have attained a higher degree of development, than chemistry had reached in their school days, they would not have placed on the pages of the annals of chemistry so glaring confirmations of the truth of Lichtenberg's keen remark which for over a century has enjoyed perennial youth:

Wer nur Chemie versteht, Versteht auch die nicht recht.
He who only understands chemistry, doesn't even understand it correctly."

96. Hinrichs, Gustavus (1864). *The Density, rotation and relative age of the planets.* New-Haven. Hinrichs observed through mathematical analyses that there were certain parallels with macrophenomenon in the universe with the infinitesimally small relationships of atoms and subatomic entities.

Travaux de 1873 à 1831 sur la rotation des molécules déterminant les états d'aggnégation de la matière, travaux de 1812 à 1912 sur la détermination des vrais poids atomiques sur le monument Lavoisier / [Gustave D. Hinrichs] / Saint-Louis [Etats-Unis] : Mo , 1912 [Works from

1873 to 1831 on the rotation of molecules determining the states of aggregation of matter, work from 1812 to 1912 on the determination of the true atomic weights on the monument Lavoisier / [Gustave D. Hinrichs] Louis [United States of America]: Mo, 1912].[Notes sur la mécanique des atomes] Série 6 : XLI-XLVIII, [Texte imprimé] : présentées par M. Marcelin Berthelot / par M. Gustave Hinrichs,... / Paris : impr. Gauthier-Villars , 1906-1907

97. Hinrichs, Gustavus Detlef (1894). *The true atomic weights of the chemical elements and the unity of matter.* (St. Louis: C. G. Hinrichs) (page images at HathiTrust; US access only)

98. Tesla, Nikola. TESLA RIDICULED EINSTEIN! Tesla disputed that the ether, does in fact exist. "You are wrong, Mr. Einstein - Aether does exist!" – "I think the material bodies do not gravitate between each other but it is the aether that makes one material body to press to another." http://www.clubconspiracy.com/forum/f30/Einstein-vs-tesla-good-article-explaining-573.html

"There is no conflict between the ideal of religion and the ideal of science, but science is opposed to theological dogmas because science is founded on fact. To me, the universe is simply a great machine which never came into being and never will end. The human being is no exception to the natural order. Man, like the universe, is a machine. Nothing enters our minds or determines our actions which is not directly or indirectly a response to stimuli beating upon our sense organs from without. Owing to the similarity of our construction and the sameness of our environment, we respond in like manner to similar stimuli, and from the concordance of our reactions, understanding is born." In the course of ages, mechanisms of infinite complexity are developed, but what we call "soul " or "spirit," is nothing more than the sum of the functionings of the body. When this functioning ceases, the "soul" or the "spirit" ceases likewise.

I expressed these ideas long before the behaviorists, led by Pavlov in Russia and by Watson in the United States, proclaimed their new psychology. This apparently mechanistic conception is not antagonistic to an ethical conception of life. The acceptance by mankind at large of these tenets will not destroy religious ideals. Today Buddhism and Christianity are the greatest religions both in number of disciples and in importance. I believe that the essence of both will be the religion of the human race in the twenty-first century." http://www.pbs.org/tesla/res/res_art11.html

99. Martinus. *Livets Bog (The Book of Life)* consists of 7 volumes and is Martinus' major, central work. Along with *The Eternal World Picture* vols. 1-5, which contains 77 symbols, it constitutes the core of his description of his spiritually scientific world picture. Nature is the "book of life", whose language the human being is learning to understand. This takes place through, among other things, the experiences of science. Without knowledge of the physical laws of Nature, people would be unable to use

the forces of Nature and create modern technology. By also acquiring knowledge of the spiritual laws of Nature, it will gradually be possible for us to understand ourselves and others, and finding meaning, logic and love in existence. http://www.martinus.dk/dk/en/martinus-writings/livets-bog/

76. Foyer, Christine H. (1984). *Photosynthesis.* New York: Wiley.

77. Govindjee, and Coleman, W. J. (1990). "How Plants Make Oxygen." *Scientific American* 262:50–59.

78. Wong, Kate (2000). "Photosynthesis's Purple Roots." *Scientific American.* Available from http://www.sciam.com .

79. Meyl, K., "Self-consistent electrodynamics," PIERS Proceedings, 172–177, Moscow, Russia, August 19–23, 2012.

80. W. Bernard Carlson (2013). *Tesla: Inventor of the Electrical Age*, Princeton University Press, page 301.

81. Küpfmüller, K. (1988). Einführung in Die Theoretische Elektrotechnik, Springer, Verl. 12, Aufl.

82. Jackson, J. D. (1975). *Classical Electrodynamics*, 2nd Edition, Wiley & Sons, NY.

83. Maxwell, J. C. (1873), *A Treatise on Electricity and Magnetism*, Dover Publications, New York.

84. Meyl, Konstatin (2012). *Scalar Wave Transponder*, 3rd Edition, Villingen-Schwenningen, INDEL Verlag.

85. Zinke, Brunswig (1986). *Lehrbuch der Hochfrequenztechnik*, Band 1, 3rd Edition, 335, SpringerVerlag.

86. Ash, D. and P. Hewitt (1990), *Science of the Gods,* Gateway Books, Bath, England.

87. Thomson, W. (1887).   Philosophical Magazine, July 1887.

88. Maxwell, J. C. (1798). *"Atoms,"* Encyclopaedia Britannica, Vol. 2, 467. 89. Thomson, J. J. (1883). *On the Motion of Vortex Rings*, London.

90. Lehner, G. (1990). *Elektromagnetische Feldtheorie*, 1st Edition, 541, Springer Verlag.

91. Morris, D. J. P., et al. (2009). *"Magnetische monopole in magnetischem Festkorper entdeckt,"* Pressemitteilung vom 3.9.09 der Hermann von Helmholtz-Gemeinschaft e.V. Berlin.

92. Meyl, K. (1994). *Dreidimensionale Nichtlineare Berechnung Von Wirbelstromkupplungen*, Diss. Universität Stuttgart, (Wirbelströme, INDEL Verlag 1991).

93. Meyl, K.(1995). *"Wirbel des elektrischen feldes,"* EMC Journal, Vol. 1, 56–59, 1995, 6. J, ISSNm 0945-3857.

94. Treskatis, T. (2010). *"Frequenzabhängigkeit der dielektrischen Verluste eines metallisierten Kunststoff Folienkondensators,"* University Konstanz.

95. Yializis, A., S. W. Cichanowski, and D. G. Shaw (1980). *"Electrode corrosion in metallized polypropylene capacitors,"* Proceedings of IEEE, International Symposium on Electrical Insulation, Boston, Mass.

96. Taylor, D. F. (1984). *"On the mechanism of aluminium corrosion in metallized film capacitors,"* IEEE Transactions on Electrical Insulation, Vol. EI-19, No. 4, 288–293.

97. Aleksandar Marinčić, Vojin Popović (1999). *Nikola Tesla.* Publisher: Zavod za udžbenike i nastavna sredstva (Institute for textbooks and teaching aids).

98. Tesla, Nikola (1992). *Nikola Tesla On His Work With Alternating Currents And Their Application To Wireless Telegraphy, Telephony, And Transmission of Power*, pp. 159-163. Edited by L.I. Anderson, Sun Publishing, Denver.

100. Gitt, Werner (2002). *Am Anfang war die Information. Herkunft des Lebens aus der Sicht der Informatik, Was ist Information?* Herkunft der Information, Naturgesetze über Information, Weitreichende Schlussfolgerungen bezüglich Menschenbild, Evolution und Urknall. 3., überarb. und erw. Aufl. Holzgerlingen: Hänssler 2002b, S. 12. 2 Ebd.

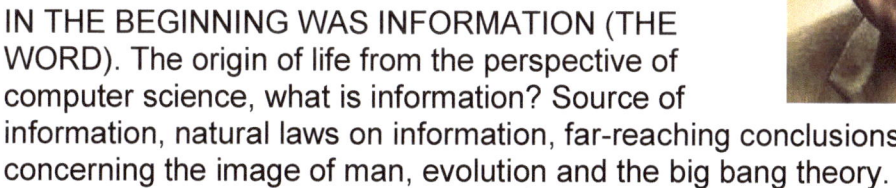

IN THE BEGINNING WAS INFORMATION (THE WORD). The origin of life from the perspective of computer science, what is information? Source of information, natural laws on information, far-reaching conclusions concerning the image of man, evolution and the big bang theory.

101. Lydia Matthes. Seminararbeit im Bereich Glaube und Wissenschaft, *Information in der DNS Evolution oder Schöpfung?*

102. Haken, Hermann u. Maria Haken-Krell (1989: Entstehung von biologischer Information und Ordnung. Darmstadt: Wiss. Buchges. (Dimensionen der modernen Biologie 3).

103. Heaviside, Oliver (1971). *Electromagnetic theory.* Appendices: D. On compressional electric or magnetic waves. Chelsea Pub Co; 3rd edition.

104. Corum, K. L., and J. F. Corum (1994). *The Zenneck surface wave*, Nikola Tesla, Lightning observations, and stationary waves, Appendix II. 1994.

105. Konstatin Meyl. *Scalar waves Advanced Concepts for Wireless Energy Transfer.*

106. Nikola Tesla (1900). *Apparatus for transmission of electrical energy*.

 US-Patent No. 645,576, N.Y. 20.3.1900.

107. Nikola Tesla (1905). *Art of transmitting electrical energy through the natural mediums*, US-Patent No. 787,412, N.Y. 18.4.1905.

108. Konstantin Meyl (1998). *Elektromagnetische Umweltverträglichkeit*, Teil 1: Umdruck zur Vorlesung, Villingen-Schwenningen 1996, 3.Aufl. 1998. Teil 2: *Energietechnisches Seminar 1998*, 3. Auflage 1999. Teil 3: Informationstechnisches Seminar 2002, auszugsweise enthalten in: K. Meyl: Skalarwellentechnik, Dokumentation für das Demonstrations-Set,INDEL-Verlag, Villingen-Schwenningen. Konstantin Meyl: *Scalar Waves*, INDEL-Verlag.

http://www.k-meyl.de

Prof. Dr.-Ing. Konstantin Meyl, 1.TZS

Leopoldstraße 1, D-78112 St. Georgen/Schwarzwald (Germany)

Tel.: +49-7724-1770, Fax.: +49-7724-9486720

Email: meyl@k-meyl.de

Internet: http://www.k-meyl.de

109. Schumann W. O. (1952). "Über die Dämpfung der elektromagnetischen Eigenschwingnugen des Systems Erde – Luft – Ionosphäre". Zeitschrift für Naturforschung A. 7: 250–252. Bibcode:1952ZNatA...7.250S. doi:10.1515/zna-1952-3-404.

110. Sheldrake, R. (1995). *Seven Experiments That Could Change the World*. (New York: Riverhead Books).
As effective scalar waves in resonance not only transmit information but also energy, even a suitable model for the phenomenon of telekinesis is found. Just as the DNA wave is radiating from a nucleus, a cell assembly, or even a human body, suitable waves can radiate in, that is, a person can absorb energy and information of people in whose aura he is, or by thinking of someone, capable of working even over long distances (Sheldrake, 1995; Engels, 2011).

111. Engels, J.W. (2011). "Distance measurements for DNA and RNA in vitro

And in vivo", Proceedings of the Second World DNA and Genome Day, China, p. 64.

112. Meyl, K. (2010). *Self-Consistent Electrodynamics. The Unified Theory is Evolving, if the Discovered Potential Vortex Replaces the Vector Potential in the Dielectric.* (Villingen: INDEL-Verlag). ISBN 978-3-940 703-15-6.

113. Meyl, K. (2011a). *DNA and Cell Resonance, Communication of Cells Explained by Field Physics Including Magnetic Scalar Waves*, 2nd edition (Villingen: INDEL Publ.) Available at www.etzs.de. ISBN 973-3-940 703-17-0. The DNA generates a longitudinal wave that propagates in the direction of the magnetic field vector. The computed frequencies from the structure of the DNA agree with those of the bio-photon radiation as predicted. The optimization of efficiency by minimizing the conduction losses leads to the double
helix structure of DNA. Scientific bases are formed on the fact that potential vortexes are an essential component of a scalar waves, as discovered in 1990. The basic approach for an extended field theory was confirmed in 2009 with the discovery of magnetic monopoles.

114. Meyl, K. (2011b). "DNA–Reading and writing by scalar waves." Proceedings of the Second World DNA Day, China, Track 2.7, p.101. Neue Horizonte in der Medizin.

115. Scalar Waves and Longitudinal Waves.
http://www.downloads.imune.net/medicalbooks/WHAT%20ARE%20SCALAR%20Waves.pdf
SW/LW in free space however are quite real. Beside Tesla, empirical work carried out by Electrical Engineers such as Eric Dollard, Konstantin Meyl, Thomas Imlauer, and Jean-Louis Naudin (to name only a few) have clearly demonstrated their existence experimentally. These waves seem able to exceed the speed of light, pass through EM shielding (aka Faraday Cages), and produce over-unity (more energy out than in) effects. They seem to propagate in a yet unacknowledged counter-spatial dimension (aka hyper-space, pre-space, false-vacuum, Aether, implicit order, etc.).

116. 10 April 2013
From: B. Rule, 3931 Brookfield Ave, Louisville, KY 40207-2001
To: RADM Richard P. Breckenridge, Deputy Chief of Naval Operations Warfare
Systems (OPNAV N97), 2000 Navy Pentagon, Washington, DC 20350-2000
Subj: Information and Security Issues Associated with the Loss of the USS THRESHER (SSN-593) on 10 April 1963 and Information on the Loss of the USS SCORPION (SSN-589) on 22 May 1968.
http://www.jag.navy.mil/library/jagman investigations.htm.

117. Eric Dollard, "Wireless Engineer" (1968). *Theory of Wireless Power.*

118. Ruth Sime (1996). *Lise Meitner: A Life in Physics*, 305.
Lise Meitner, wife and major contributor to the work of Albert Einstein, was one of the few scientists that courageously opposed the development of the atomic bomb on moral grounds. In response to her invitation in 1943 to work with Otto Robert Frisch and some British scientists at Los Alamos during the Manhattan Project she stated: "I will have nothing to do with a bomb!"

119. Tesla's Death Rays, extracted from "The Fantastic Inventions of Nikola Tesla", at the Scribd Website. On July 11, 1934, the New York Times ran a story which was headlined: TESLA AT 78 BARES NEW "DEATH-BEAM." Tesla claimed to have invented a "defensive" weapon powerful to destroy 10,000 airplanes at a distance of 250 miles away. The scientist tells of an apparatus that he says will kill without a trace.

http://www.bibliotecapleyades.net/tesla/esp_tesla_33.htm

120. T.E. Bearden and Floyd Sweet (1991). *UTILIZING SCALAR ELECTROMAGNETICS TO TAP VACUUM ENERGY* [Abridged].
Summary:
http://homepages.ihug.co.nz/~sai/Beard_scal_vac.html
Long ago, Albert Einstein said these words: "It would of course be a great step forward if we succeeded in combining the gravitational field and the electromagnetic field into a single structure. Only so could the era in theoretical physics inaugurated by Faraday and Clerk Maxwell be brought to a satisfactory close." And Teilhard de Chardin wrote: "Someday, after we have mastered the winds, the waves, the tides and gravity, we shall harness for God the energies of love. Then for the second time in the history of the world man will have discovered fire." The authors fervently believe they have come upon fire for the second time, as allegorized by de Chardin. If so, let us all use the knowledge wisely.
Floyd Sweet- Association of Distinguished American Scientists 2311 Big Cove Road Huntsville, Alabama 358010
T. E. Bearden - Association of Distinguished American Scientists 2311 Big Cove Road Huntsville, Alabama 35801

121. Planck, Max. (1932). *Where is Science Going?* 168.

122. Johannes Keppler (1599). "Those laws [of nature] are within the grasp of the human mind; God wanted us to recognize them by creating us after his own image so that we could share in his own thoughts."

Letter (9/10 Apr 1599) to the Bavarian chancellor Herwart von Hohenburg. Collected in Carola Baumgardt and Jamie Callan, *Johannes Kepler Life and Letters* (1953), 50.

123. W.O. Schumann (1952). The Earth behaves like an enormous electric circuit. Given that the earth's atmosphere carries a charge, a current and a voltage, it is not surprising to find such electromagnetic waves. The Schumann resonance facilitates the natural resonance in the human body and helps to recover autonomic nerve balance, remove stress, and intensify concentration. The resonant properties of this terrestrial cavity were first predicted by the German physicist W. O. Schumann between 1952 and 1957, and first detected by Schumann and König in 1954. The first spectral representation of this phenomenon was prepared by Balser and Wagner in 1960.

The Schumann Resonances are quasi standing wave electromagnetic waves that exist in this cavity. Like waves on a spring, they are not present all the time, but must be 'excited' to be observed. Schumann predicted that there are electromagnetic standing waves in the atmosphere, within the cavity formed by the surface of the Earth and the ionosphere. The limited dimensions of the earth and the conductive ionosphere cause this waveguide to act as a resonant cavity for electromagnetic waves in the ELF (extremely low frequency) band. This cavity is naturally excited by energy from lightning strikes. They are not caused by anything internal to the Earth, its crust or its core. They seem to be related to electrical activity in the atmosphere, particularly during times of intense lightning activity. They occur at several frequencies between 6 and 50 cycles per second; specifically, 7.8, 14, 20, 26, 33, 39 and 45 Hertz, with a daily variation of about +/- 0.5 Hertz.

Die Schumann-Resonanzen sind quasi stehende Welle elektromagnetische Wellen die in diesem Hohlraum vorhanden ist. Wie Wellen auf einer Feder, sind sie nicht vorhanden die ganze Zeit, sondern muss "angeregt" werden um zu beobachten. Schumann vorhergesagt, dass es elektromagnetische stehende Wellen in der Atmosphäre, innerhalb der von der Oberfläche der Erde gebildeten Hohlraum und der Ionosphäre gibt. Die begrenzten Abmessungen der Erde und die leitende Ionosphäre verursachen dieser Wellenleiter als Resonanzraum für elektromagnetische Wellen im ELF (extrem niedrige Frequenz) Band zu dienen . Dieser Hohlraum wird durch Energie aus Blitzeinschlägen natürlich erregt. Sie werden nicht durch irgendetwas im Inneren der Erde, ihre Kruste oder dessen Kern verursacht. Sie scheinen von elektrischer Aktivität in der Atmosphäre verursacht, besonders in Zeiten intensiver Blitzaktivität . Sie treten auf mehreren Frequenzen zwischen 6 und 50 Zyklen pro Sekunde; g'-0p[;\-0enauer gesagt, 7,8; 14; 20; 26; 33; 39;und 45 Hertz, mit einer täglichen Variation von etwa +/- 0,5 Hertz.

So long as the properties of Earth's electromagnetic cavity remains about the same, these frequencies remain the same. Presumably there is some change due to the solar sunspot cycle as the Earth's ionosphere changes in response to the 11-year cycle of solar activity. Schumann resonances are most easily seen between 2000 and 2200 UT. Much of the research in the last 20 years has been conducted by the Department of the Navy which investigates Extremely Low Frequency communication with submarines.

For medical purposes and diagnosis, the Schumann resonance facilitates the natural resonance in the human body and helps to recover autonomic nerve balance, remove stress, and intensify concentration. The human brain produces electromagnetic waves, which lie in the range between 1 and 40 Hz. This spectrum is divided into four ranges in medical analysis, which are associated with different consciousness conditions:

1. Delta waves (1 - 3 Hz) which characterize dreamless deep sleep and koma conditions.
2. Theta waves (4 - 7 Hz) which are characteristic for deep sleep.
3. Alpha waves (8 – 12 Hz) which occur in the relaxed awake condition, e.g. in meditation or briefly before falling asleep, and\or immediately after awaking.
4. Beta waves (13 -40 Hz) which are dominant in the normal awake condition.

For more information, see:

"Handbook of Atmospheric Electrodynamics, vol. I", by Hans Volland, 1995
published by the CRC Press. Chapter 11 is entirely on Schumann Resonances and is written by Davis Campbell at the Geophysical Institute, University of
Alaska, Fairbanks AK, 99775.

"The atmosphere is a weak conductor and if there were no sources of charge, its existing electric charge would diffuse away in about 10 minutes. There is a 'cavity 'defined by the surface of the Earth and the inner edge of the ionosphere 55 kilometers up. At any moment, the total charge residing in this cavity is 500,000 Coulombs. There is a vertical current flow between the ground and the ionosphere of 1 - 3 x 10^-12 Amperes per square meter. The resistance of the atmosphere is 200 Ohms. The voltage potential is 200,000 Volts. There are about 1000 lightning storms at any given moment worldwide. Each produces 0.5 to 1 Ampere and these collectively account for the measured current flow in the Earth's 'electromagnetic' cavity."
http://image.gsfc.nasa.gov/poetry/ask/q768.html

Appendix I

DNA molecules can 'teleport', Nobel Prize winner claims

Luc Montagnier, a Nobel Prize winning biologist, has ignited controversy after publishing details of an experiment in which a fragment of DNA appeared to 'teleport' or imprint itself between test tubes.

Luc Montagnier's team, surrounded two test tubes with a weak electromagnetic field of 7 Hz, one of which contained a tiny piece of bacterial DNA, and the other pure water. The DNA was amplified using a polymerase chain reaction, and eighteen hours later, the DNA was detected in the test tube which only contained pure water. This phenomenon had not been detected previously, presumably because the DNA sample had to diluted many times over for the experiment to work. Since the bases projected or imprinted themselves across space, rather than simply moving from one place to another, Montagnier's team has given evidence of "teleportation" of DNA information.

The apparent imprinting of the DNA on the water, a quantum effect, is normally assumed to occur in imperceptible fractions of a second, but not minutes and hours. In addition, quantum effects usually result at very low temperatures near absolute zero (Absolute zero is the lower limit of the thermodynamic temperature scale, a state at which the enthalpy and entropy of a cooled ideal gas reaches its minimum value, taken as 0. The theoretical temperature is determined by extrapolating the ideal gas law; by international agreement, absolute zero is taken as −273.15° on the Celsius scale, equal to −459.67° on the Fahrenheit scale). What is startling about this biological process is that the underlying "quantumness" of the effect has occurred at room temperature.

 "It is hard to understand how the information can be stored within water over a timescale longer than picoseconds," said the Ruhr University in Bochum's Klaus Gerwert, who was quoted in the *New Scientist* magazine. https://www.newscientist.com/article/mg20927952.900-scorn-over-claim-of-teleported-dna?

Montagnier has met much skepticism regarding his experimental results, which could "shake the foundations of several fields of science" if further corroborated by other researchers. Theoretical chemist Jeff Reimers at the University of Sydney, Australia, stated "these would be the most significant experiments performed in the past 90 years, demanding re-evaluation of the whole conceptual framework of modern chemistry." As with all scientific advances, there is a common pattern in the acceptance of new ideas:

The central role of the "little-understood quantum properties of the water molecule and not just its more obvious chemical bonding properties" in the bio-engineering of life forms suggest that "water might be a good medium in which DNA can copy itself using processes" of quantum entanglement and "teleportation." Using "quantum field theory", Montagnier has also investigated the propagation of diseases.

http://news.techworld.com/personal-tech/3256631/dna-molecules-can-teleport-nobel-prize-winner-claims/

Den centrale rolle den "lille-forståede kvanteegenskaber af vandmolekylet og ikke kun sine mere indlysende kemiske bindingsegenskaber" i bio-engineering af livsformer tyder på, at "vand kan være en god medium i hvilke DNA kan kopiere sig selv ved hjælp af processer af kvante entanglement og teleportation. Brug af kvantefeltteorien", har Montagnier også undersøgt udbredelsen af sygdomme.

Scalar Waves Kill Cancer Cells, sensational progress in cancer treatment

Saturday, 24 August 2013 14:52

Prof. Konstantin Meyl is the world's leading physicist in the research on scalar waves, both in theory and practice. There are three kinds of waves:

1. **electromagnetic waves** (Heinrich Hertz), well known to everybody

2. **electric scalar waves** (Nikola Tesla), only known by secret services and enlightened people, even called Tesla waves, and

3. **magnetic scalar wave** (Konstantin Meyl), unknown

Meyl figured out, that the **magnetic scalar wave is biological relevant,** not the electromagnetic wave. Therefore cell communication is done by magnetic scalar waves. It was detected, that cancer cells mostly communicate with their own kind, not with healthy cells. So Meyl poisoned cancer cells and while dying, they send a special signal received only by other cancer cells, which causes them to die also. Healthy cells do not receive this signal, which is transmitted by magnetic scalar waves.

Konstantin Meyl är världens ledande fysikprofessor på området skalärvågor. Det finns tre sorters vågor:

1. elektromagnetiska vågor (Heinrich Hertz), välkända av alla

2. elektriska skalärvågor (Nikola Tesla), enbart kända hos folk från säkerhetstjänsterna och upplysta människor

3. magnetiska skalärvågor (Konstantin Meyl) okända

Meyl listade ut att det är de magnetiska skalärvågorna, som är biologiskt relevanta, inte de elektromagnetiska. Därför sker informationsutbytet mellan cellerna med hjälp av magnetiska skalärvågor. Man har upptäckt, att cancerceller mest kommunicerar med sina egna, inte med friska celler. Därför förgiftade Meyl cancerceller och medan de dog sände de en speciell signal som enbart mottogs av cancerceller, vilket ledde till deras död. Sunda celler mottog icke denna signal, som transporterades med magnetiska skalärvågor.

With this method, only cancer cells are targeted. That is a huge difference to the chemotherapy from today, which target almost the whole body and often kill the patients slowly.

White TV wants to point out, that the killing of cancer cells with scalar waves is better than chemo, but not the optimum, because to attack

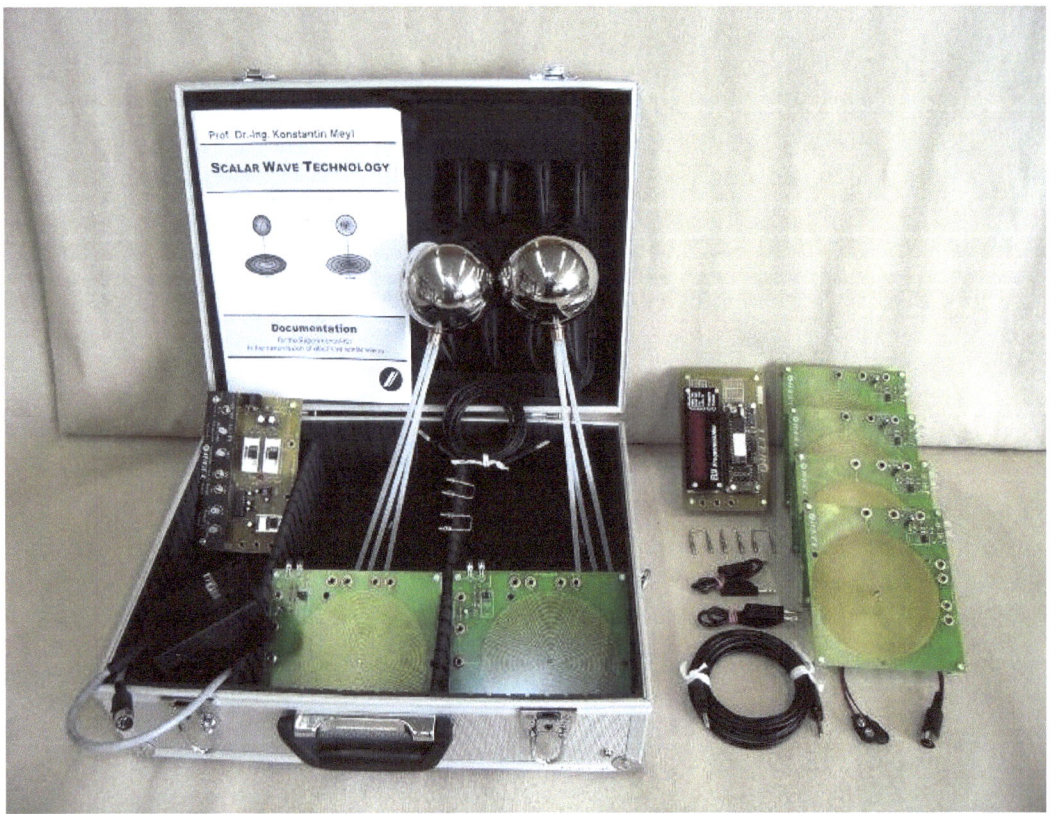

cancer cells and tumors is only to attack the symptoms but not the cause of cancer. Here other methods must be used.

Med denna metod attackeras enbart cancerceller. Detta utgör en stor skillnad mot dagens kemoterapi, som angriper nästan hela kroppen och låter folk ofta dö på ett plågsamt sätt.

White TV vill dock påpeka att dödandet av cancerceller med skalärvågor visserligen är bättre än dagens kemo, men med denna metod attackeras enbart symptomen för cancern, inte orsaken. I den kampen bör andra metoder användas.

Source: Scalar Waves Kill Cancer Cells from Henning Witte on Vimeo.

Appendix III

The Connection between the Schumann Resonance of the Earth and the Alpha Rhythm of Human Brainwaves

The transmission of longitudinal electric waves. Like Meyl, the famous experimental physicist Tesla had conducted a historical experiment 100 years ago, and measured the properties of longitudinal electrical waves. He patented his findings in the wireless transmission of energy in 1900. <u>Tesla also discovered that more energy arrived at the receiver than was supplied to the transmitter in his experimental setup, and called it a "magnifying transmitter."</u>

By the effect back on the transmitter Tesla observed that the resonance of the earth was 12 Hz. Since the Schumann resonance of a wave, traveling at the speed of light is 7.8 Hz, Tesla concluded that his longitudinal electrical wave had a traveled at 1.5 times the speed of light, therefore exceeding Einstein's calculations.

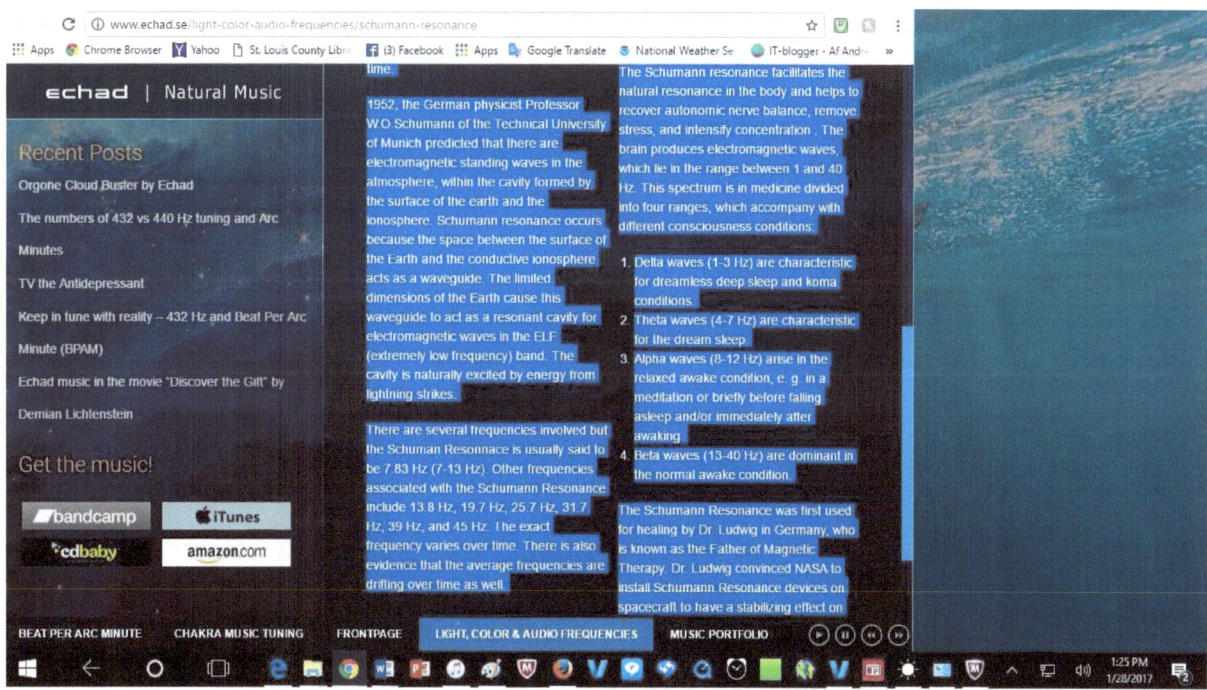

At the time when Schumann published his research results in the journal *Technische Physik*, Dr Ankermueller, a physician, immediately made **the connection between the Schumann resonance and the alpha rhythm of brainwaves.** He found the thought of the earth having the same natural resonance as the brain very exciting and contacted Professor Schumann, who in turn asked a doctorate candidate to research this phenomenon. This candidate was Herbert König who became Schumann's successor at Munich University. König demonstrated a correlation between Schumann Resonances and brain rhythms. He compared human EEG recordings with natural electromagnetic fields of the environment (1979) and found that the main frequency produced by Schumann oscillations is very close to the frequency of alpha rhythms.   http://www.earthbreathing.co.uk/sr.htm

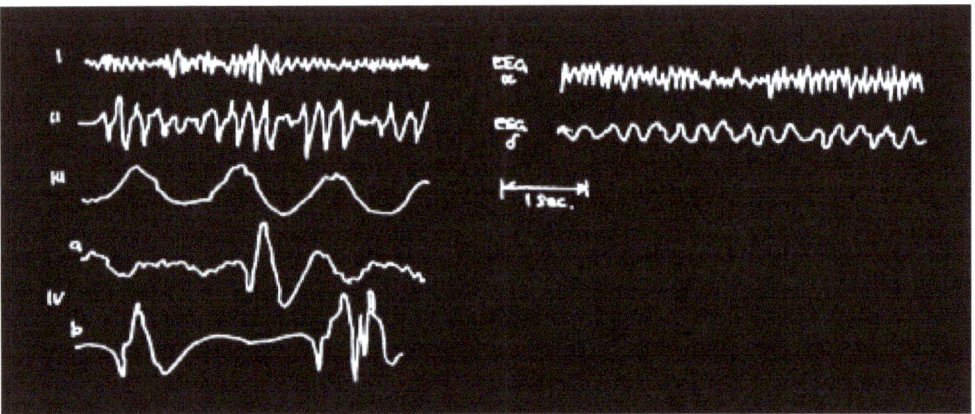

Natural electromagnetic processes in the environment (I-IV), human EEG readings in comparison. Schumann oscillations (I) and the EEG a-rhythm, as well as locally conditioned fluctuations of the electric field (II) and the EEG d-rhythm, show a noticeable similarity in their temporal variation. From König, 1979.

Lightning discharges are considered to be the primary natural source of Schumann resonance excitation; lightning channels behave like huge antennas that radiate electromagnetic energy at frequencies below about 100 kHz. These signals are very weak at large distances from the lightning source, but the Earth–ionosphere waveguide behaves like a resonator at ELF frequencies and amplifies the spectral signals from lightning at the resonance frequencies.

In an ideal cavity, the resonant frequency of the n-th mode f_n is determined by the Earth radius a and the speed of light c.

$$f_n = \frac{c}{2\pi a} \sqrt{n(n+1)}$$

The real Earth–ionosphere waveguide is not a perfect electromagnetic resonant cavity. Losses due to finite ionosphere electrical conductivity lower the propagation speed of electromagnetic signals in the cavity, resulting in a resonance frequency that is lower than would be expected in an ideal case, and the observed peaks are wide. In addition, there are a number of horizontal asymmetries – day-night difference in the height of the ionosphere, latitudinal changes in the Earth's magnetic field, sudden ionospheric disturbances, polar cap absorption, variation in the Earth radius of ± 11 km from equator to geographic poles, etc. that produce other effects in the Schumann resonance power spectra.

"Professor R. Wever from the Max Planck Institute for Behavioural Physiology in Erling-Andechs, built an underground bunker which completely screened out magnetic fields. Student volunteers lived there for four weeks in this hermetically sealed environment. Professor Wever noted that the student's circadian rhythms diverged and that they suffered emotional distress and migraine headaches. As they were young and healthy, no serious health conditions arose, which would not have been the case with older people or people with a compromised immune system. After only a brief exposure to 7.8 Hz (the very frequency which had been screened out), the volunteers' health stabilized again."

With the advent of new wireless technology, microwaves pulsed at frequencies close to Schumann Resonance as in mobile telephony, another threat is emerging. We may be creating an environment that is literally `out of tune' with Nature itself. **There is an urgent need for us to understand how everything alive responds to the subtlest changes in magnetic and electromagnetic fields surrounding us.** For instance, we need to examine the possible interaction between magnetite crystals within cells and manmade magnetic fields in the environment.

http://www.earthbreathing.co.uk/sr.htm

Konstatin Meyl describes in his treatise, *DNA and Cell Resonance: Magnetic Waves Enable Cell Communication,* how two cells communicate with each other where the magnetic scalar wave uses the genetic code chemically stored in the base pairs of the genes and electrically modulates them.

In the beginning was the information. Origin of life from the perspective of computer science, what is information? Source of information, natural laws on information, far-reaching conclusions concerning the image of man, evolution and big bang that propagates in the direction of the magnetic field vector. Computed frequencies from the structure of DNA agree with those of the predicted biophoton radiation. The optimization of efficiency by minimizing the conduction losses leads to the double-helix structure of DNA. The vortex model of the magnetic scalar wave not only covers many observed structures within the nucleus perfectly, but also explains the hyperboloid channels in the matrix when two cells communicate with each other. Potential vortexes are an essential component of a scalar waves, as discovered in 1990. **The basic approach for an extended field theory was confirmed in 2009 with the discovery of magnetic monopoles.** For the first time, this provides the opportunity to explain the physical basis of life not only from the biological discipline. Nature covers the whole spectrum of known scientific fields of research, and interdisciplinary understanding is required to explain its complex relationships. The characteristics of the potential vortex are significant. With its concentration effect, it provides for miniaturization down to a few nanometers, which allows enormously high information density in the nucleus.

The magnetic scalar wave is "suitable to use genetic code chemically stored in the base pairs of the genes and electrically modulate them." It then "piggybacks'" information from the cell nucleus to another cell. At the receiving end, the reverse process takes place and the transported information is converted back into a chemical structure. The necessary energy required to power the chemical process is provided by the magnetic scalar wave itself.

http://www.k-meyl.de/go/Primaerliteratur/Magnetic_Waves-Enable-Cell_Communication.pdf

As effective scalar waves in resonance not only transmit information but also energy, even a suitable model for the phenomenon of telekinesis is found. Just as the DNA wave radiates from a nucleus, a cell assembly, or even a human body, suitable waves can radiate in, that is, a person can absorb energy and information of people in whose aura he is, or by thinking of someone, capable of working even over long distances. (Sheldrake, 1995; Engels, 2011).

No chemical process is more important to life on Earth than photosynthesis —the series of chemical reactions that allow plants to harvest sunlight and create carbohydrate molecules. Without photosynthesis, not only would there be no plants, the planet could not sustain life of any kind. In plants, photosynthesis occurs in the thykaloid membrane system of chloroplasts. Many of the enzymes that allow photosynthesis to occur are transmembrane proteins embedded in the thykaloid membranes. The chemistry involved is described at the following website.

http://www.chemistryexplained.com/Ny-Pi/Photosynthesis.html

The most basic summary of the photosynthesis process can be shown with a net chemical equation:

$$6CO_2 (g) + 6 H_2O(l) + h\nu \rightarrow C_6H_{12}O_6 (s) + 6O_2 (g)$$

The symbol hv is used to depict the energy input from light (in the case of most plants, sunlight). This chemical equation, however, is a dramatic simplification of the very complicated series of chemical reactions that photo-synthesis involves. It also implies that the only product is glucose, $C_6H_{12}O_6$ (s), which is also a simplification...

The critical step of the light cycle is the absorption of electromagnetic radiation by a pigment molecule. The most famous pigment is chlorophyll, but other molecules, such

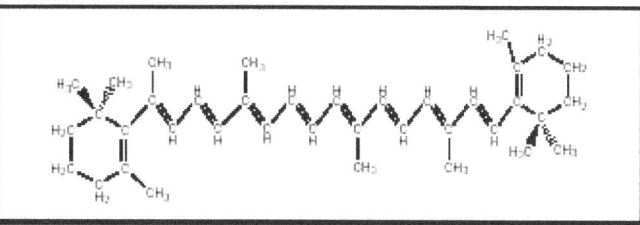

as β- carotene, also absorb light. Together, these pigment molecules form a type of light harvesting antennae that is more efficient at interacting with sunlight than would be possible with the pigments acting alone. When the light is absorbed, electrons in the pigment molecule are excited to high energy states. A series of enzymes called electron transport systems help channel the energy present in these electrons into reactions that store it in chemical bonds.

http://www.chemistryexplained.com/Ny-Pi/Photosynthesis.html#ixzz4WbdlLwxK

http://www.chemistryexplained.com/Ny-Pi/Photosynthesis.html#ixzz4WbcpPt57

APPENDIX IV

BIOGRAPHY OF HANS CHRISTIAN ØRSTED

Hans Christian Ørsted. född den 14 augusti 1777 i Rudkøbing på Langeland, död den 9 mars 1851 i Köpenhamn, var en dansk fysiker och kemist, bror till statsminister Anders Sandøe Ørsted och farbror till botanikern med samma namn. Han är känd för att ha varit den förste att framställa aluminium 1825 och för att 1820 ha upptäckt sambandet mellan elektricitet och magnetism. Han betraktas som en av den danska guldålderns tongivande personligheter.

Hans Christian Ørsted, born on August 14, 1777 in Rudkøbing on Langeland, died March 9, 1851 in Copenhagen, was a Danish physicist and chemist, the brother of Prime Minister Anders Sandøe Ørsted and uncle to the botanist of the same name. He is known for having been the first to produce aluminum in 1825 and 1820 to have discovered the link between electricity and magnetism. He is regarded as one of the influential personalities of the Danish Golden Age.

Ørsteds far var apotekare, och Hans Christian fick hemundervisning. Han avlade studentexamen 1794 i Köpenhamn efter endast ett halvårs studier. År 1797 tog han farmaceutisk examen. Han var sedan under ett par år föreståndare för ett apotek i Köpenhamn, medan han fortsatte med studier i filosofi, estetik, fysik och kemi. Han fick universitetets guldmedalj för besvarandet av en prisfråga i estetik 1797 och i medicin 1798. Han blev filosofie magister på en avhandling om Kants naturfilosofi 1799 och anställdes 1800 som adjunkt vid den medicinska fakulteten. Ørsted reste runt i Europa och återvände 1803 till Köpenhamn, där han försörjde sig som privatdocent. Hans föreläsningar var populära och han utnämndes till professor i Köpenhamn 1806. Ørsted var en av grundarna av Polyteknisk Læreanstalt och var dess förste direktör 1829-1851.

Ørsted's father was a pharmacist, and Hans Christian was home schooled (It does work!) He graduated with a baccalaureate in 1794 in Copenhagen after only half a year of studies. In 1797 he took the pharmaceutical degree. He was then for a few years the director of a pharmacy in Copenhagen, while he continued with studies in philosophy, aesthetics, physics and chemistry. He received the University Gold Medal for the answer to a prize question of aesthetics in 1797, and in medicine in 1798. He received a Master of Arts degree with a thesis on Kant's natural philosophy in 1799. He was hired in 1800 as a lecturer at the Faculty of Medicine. Ørsted traveled around Europe and returned in 1803 to Copenhagen, where he supported himself as a private docent. His lectures were popular and he was appointed professor in Copenhagen in 1806. Ørsted was one of the founders of the Polytechnic Læreanstalt and was its first director from 1829 to 1851.

Ørsted Vater war Apotheker, und Hans Christian bekam Hausunterricht . Er verdiente Abiturs 1794 in Kopenhagen nach nur einem halben Jahr von Studien. Im Jahr 1797 nahm er den pharmazeutischen Grad. Er war dann in ein paar Jahren Leiter einer Apotheke in Kopenhagen, wie er mit dem Studium der Philosophie, der Ästhetik, Physik und Chemie fortgesetzt. Er erhielt die Universität Goldmedaille für die Antwort auf eine Preisfrage der Ästhetik im Jahre 1797 und in der Medizin im Jahr 1798. Er wurde Master of Arts mit einer Arbeit über Kants Naturphilosophie im Jahre 1799 und wurde im Jahr 1800 als Dozent an der Fakultät für Medizin eingestellt. Ørsted reiste um Europa und kehrte nach Kopenhagen im Jahre 1803, wo er sich als Privatdozent unterstützt. Seine Vorlesungen waren sehr beliebt und wurde er zum Professor in Kopenhagen im Jahre 1806 Ørsted war einer der Gründer der Polytechnischen Læreanstalt und war ihr erster Direktor 1829-1851

Även inom kemin blev Ørsted framgångsrik; han var till exempel den förste att utvinna aluminium ur kaliumamalgam och aluminiumklorid. Han var också aktiv i strävan att främja naturvetenskapen i Danmark och grundade därför *Selskabet for Naturlærens Udbredelse*. Han invaldes 1822 som utländsk ledamot nummer 242 av svenska Kungliga Vetenskapsakademien. Han var vän till Hans Christian Andersen.[1]

Auch in der Chemie war Oersted erfolgreich; Er war zum Beispiel der erste der Aluminium aus Kaliumamalgam und Aluminiumchlorid extrahiert. Er war auch bei den Bemühungen aktiv Wissenschaft in Dänemark zu fördern und gründete deshalb *Selskabets für Naturlærens Udbredelse*. Er wurde im Jahre 1822 als auswärtiges Mitglied Nummer 242 der *Schwedischen Königlichen Akademie der Wissenschaften* gewählt. Er war ein Freund von Hans Christian Andersen.

Selv i kemi var Ørsted en succes; Han var, for eksempel, den første person at udtrække aluminium fra kalium amalgam og aluminiumchlorid. Han var også aktiv i bestræbelserne på at fremme videnskab i Danmark og grundlagde derfor *Selskabet for Naturlærens Udbredelse*. Han blev valgt i 1822 som en udenlandsk medlem nummer 242 af den *Svenske Kongelige Videnskabsakademi*. Han var en ven af HC Andersen.

H.C. Ørsted är begravd på Assistens Kirkegård i Köpenhamn. Han har fått en mätenhet uppkallad efter sig med beteckningen Oe för magnetisk fältstyrka.

H.C. Ørsted er begravet i Assistens Kirkegård i København. Han har modtaget en enhed opkaldt efter ham med navnet Oe af magnetisk feltstyrke.

H.C. Ørsted is buried in the Assistant's Cemetery in Copenhagen. He has received a unit named after him with the name Oe designating magnetic field strength.

Vid iordningställandet av ett demonstrationsförsök för en lektion 21 april 1820 lade Ørsted märke till att en kompassnål rörde sig när han slöt en strömkrets. Han började med systematiska experiment och publicerade sina resultat tre månader senare. Han försökte olika ledare, han såg att kompassnålens utvikelse var åt andra hållet när han bytte strömriktning och han visade att effekten inte kunde skärmas av trä eller glas.

Ved udarbejdelse af en demonstration for en lektion April 21, 1820, Ørsted bemærket, at en kompasnål flyttede, da han lavede et kredsløb. Han begyndte med systematiske eksperimenter og offentliggjort deres resultater tre måneder senere. Han prøvede forskellige ledere, han så, at den kompasnålen fold ellers var den anden vej, når han skiftede strømretning og han viste, at effekten ikke kunne afskærmet af træ eller glas.

Bei der Herstellung eines Demonstrations Versuch für eine Lektion 21. April 1820, fiel Ørsted ein, dass eine Kompassnadel bewegt wird, wenn er eine Schaltung gemacht. Er begann mit systematischen Experimenten und drei Monate später veröffentlichtet die Ergebnisse. Er versuchte verschiedene Stromleiter, und sah, dass die Kompassnadel fielten in die andere Richtung, als er die Stromrichtung eingeschaltet hat, und er zeigte, dass die Wirkung nicht durch Holz oder Glas abgeschirmt werden könnten.

Upptäckten fick omedelbart genomslag i vetenskapliga kretsar. Efter några månader publicerade den franska fysikern André Ampère en matematisk beskrivning av de magnetiska krafterna mellan strömförande ledare. 1824 konstruerade William Sturgeon den första elektromagneten.

Ørsted's discovery had an immediate impact in the scientific community. A few months, the French physicist Andre Ampere published a mathematical description of the magnetic forces between current-carrying conductor. In 1824 William Sturgeon constructed the first electromagnet.

I historiska sammanhang förekommer det ibland hänsyftningar på att en italiensk jurist och amatörfysiker, Gian Domenico Romagnosi, skulle ha varit en föregångare till Ørsted genom att redan 1802 påvisa och publicera en samverkan mellan magnetiska och elektriska krafter. **Romagnosis experiment reagerade emellertid bara på elektrostatiska krafter.** Ingen galvanisk ström förekom i experimentet och alltså inte heller någon elektromagnetism.

In historical context, there are sometimes references to an Italian lawyer and amateur physicist, Gian Domenico Romagnosi, would have been a precursor to Ørsted already in 1802. He already demonstrated and published an interaction between magnetic and electrical forces. Romagnosis experiments only involved electrostatic forces, however. No galvanic current existed in his experiment and therefore no electromagnetism was present.

I historisk sammenhæng, der var nogle gange henvisninger til en italiensk advokat og amatør fysiker Gian Domenico Romagnosi, han ville have været en forløber for Ørsted allerede i 1802. Han demonstreret og offentliggøret et samspil mellem de magnetiske og elektriske kræfter. Romagnosis eksperimenter reagerede imidlertid kun på elektrostatiske kræfter. Ingen galvanisk strøm eksisterede i eksperimentet, og derfor heller ikke nogen elektromagnetisme.

www.ingramcontent.com/pod-product-compliance
Lightning Source LLC
Chambersburg PA
CBHW051018180526
45172CB00002B/392